PROJECT AIR FORCE

T0122959

Robust and Resilient Logistics Operations in a Degraded Information Environment

Don Snyder, Elizabeth Bodine-Baron, Mahyar A. Amouzegar, Kristin F. Lynch, Mary Lee, John G. Drew

Prepared for the United States Air Force
Approved for public release; distribution unlimited

For more information on this publication, visit www.rand.org/t/RR2015

Library of Congress Cataloging-in-Publication Data is available for this publication.
ISBN: 978-0-8330-9830-6

Published by the RAND Corporation, Santa Monica, Calif.
© Copyright 2017 RAND Corporation
RAND® is a registered trademark.

Support RAND
Make a tax-deductible charitable contribution at
www.rand.org/giving/contribute

www.rand.org

Preface

Logistics operations depend on accurate information. If acted upon, inaccurate information can cause support operations to work counter to desired outcomes and potentially negatively affect combat operations. Even relatively small errors in support systems can, in some circumstances, have large effects on operations. Yet errors are inevitable, so logistics operations should be robust to errors, whether they are a random occurrence or the result of a deliberate, targeted cyber attack.

The U.S. Air Force asked RAND Project AIR FORCE to determine where it is most fruitful to focus effort in making changes to tactics, techniques, and procedures to improve an airman's ability to detect, evaluate, and mitigate significant corruption of logistics data. The goal is to respond to errors in data before they have a significant negative effect on combat operations.

This work was completed as part of the fiscal year 2016 project entitled "Logistics Ability to Survive and Operate After Cyber Attack" and was sponsored by the Director of Resource Integration under the Air Force Deputy Chief of Staff, Logistics, Engineering, and Force Protection and co-sponsored by the Director, Logistics, Engineering, and Force Protection, Air Force Global Strike Command. The work was conducted within the Resource Management Program of RAND Project AIR FORCE. It should be of interest to the logistics and operational communities throughout the Air Force.

RAND Project AIR FORCE

RAND Project AIR FORCE (PAF), a division of the RAND Corporation, is the U.S. Air Force's federally funded research and development center for studies and analyses. PAF provides the Air Force with independent analyses of policy alternatives affecting the development, employment, combat readiness, and support of current and future air, space, and cyber forces. Research is conducted in four programs: Force Modernization and Employment; Manpower, Personnel, and Training; Resource Management; and Strategy and Doctrine. The research reported here was prepared under contract FA7014-16-D-1000.

Additional information about PAF is available on our website: www.rand.org/paf

Contents

Figures and Tables

Summary

To safeguard the ability of the U.S. Air Force to project air power, logistics support needs to be robust and resilient against a range of threats, including cyber attack.[1] One type of cyber attack is the deliberate corruption of data. When data are corrupted, actions based on those data can hamper logistics operations, which can, in turn, negatively affect combat operations. Data corruption events are not only a clear concern for targeted attacks, but also for some routine errors.

In 2015, as part of an initiative called Logistics Ability to Survive and Operate, Air Force logistics leadership described three steps for responding to a suspected data corruption incident: (1) detect (and report) any suspected errors; (2) evaluate whether these are indeed errors, and, if so, determine whether they are merely mistakes or the result of a deliberate attack; and (3) mitigate any problems to logistics operations created by the real errors. The objective of this report is to improve the robustness and resiliency of logistics operations in the event of significant data corruption. This report examines what the logistics community can do to better detect, evaluate, report, and prioritize the response to corrupt data in order to satisfactorily continue operations. It does not discuss measures that the information security community can take or specific mitigation measures for cyber attacks against the logistics community.

Detecting

After examining the necessary steps that must occur for successful detection within the framework of high resiliency, we identified five critical areas that show the most promise to enhance the ability to detect corrupted data:

1. The detection must be sufficiently prompt. Promptness results from a combination of individuals detecting and reporting corrupted data quickly and detecting corrupted data early in the chain of custody.
2. Highly automated processes—in which humans do not see the data during normal operations—present a significant challenge for detection. Automation requires special mechanisms to assist in detecting corruption.
3. Detection of corrupted data is most critical during wartime, yet anomalies are less evident during wartime than during peacetime because wartime itself is an anomaly. Mechanisms are needed to adjust detection mechanisms from peacetime to wartime conditions.
4. For workers (airmen, civilians, and contractors) to detect anomalous data, they all need to be trained to understand the expected baseline and need to be continuously vigilant when examining data.

---

[1] We use the term *robust* to mean the ability to absorb an attack and maintain an acceptable level of function and the term *resilient* to mean the ability to recover function sufficiently rapidly after an attack.

5. Leadership must create an environment that encourages workers to report suspected anomalous data.

To address these challenges, we make three key recommendations:

First, we recommend defining, within logistics policy, what measures the logistics community should take in response to each information operations condition (INFOCON) level.[2] Of all the recommendations in this report, we assess that this action is most likely to yield positive, tangible results with the least expenditure of resources. These measures would be aimed more at operational continuity of logistics functions than at information assurance. Procedures within logistics would be specified to parallel existing procedures specified for information technology systems. For each INFOCON level, specific proactive measures would be defined in each functional area for increasing robustness and resiliency. These measures would progress with increasing threat so as to match the expended resources (and any costs to efficiency) with the threat environment. Just as in force protection condition levels, the measures to be taken would be specific and tailored, and they would be written into U.S. Air Force Instructions.

Second, we recommend considering the adoption of simple methods to detect data corruption in fully automated processes (in which humans do not "see" the data during normal operations) that are routinely used in fraud detection. Benford's Law (detailed in Chapter Three) is one relatively easy check on data, but other standard methods could also be used. The extent of use of these techniques should be set according to the cyber threat condition (INFOCON) to economize on workload. These tools are not by themselves a complete solution; rather, they can be used to flag potentially problematic data for further human review.

Third, we recommend enhancing the education and training of all members of the enterprise about the threat to operations from data corruption (and cyber attack more generally). The key objectives of such training are as follows:

- Educate workers at all levels that *all* data systems are susceptible to data corruption and cyber attack, and, therefore, vigilance is required for *all* systems and processes.
- Instill a culture in the workforce that everyone in the logistics enterprise has equity in cybersecurity, not just those with specific cybersecurity specialties.

Each level of leadership (command and supervisory) should play a role in this messaging. Senior levels (generally flag- and field-grade officers) should set the reporting of cybersecurity concerns as a priority. They should make that priority concrete by extending protection from retaliation to reporters. And they should write policy to ensure that reporting is easy to do and that guidance is specific. The goal is to shift the culture so that workers are as vigilant in reporting suspected corrupted data and other cyber concerns as they are with foreign object

---

[2] INFOCON is soon to be replaced by a cyber conditions (CYBERCON) system. INFOCON levels are global and set by the Commander, U.S. Strategic Command. However, commanders at all levels can set more-strict INFOCON levels locally for systems under their command (Chairman of the Joint Chiefs of Staff Instruction 6510.01F, February 9, 2011, directive current as of June 9, 2015; Strategic Command Directive 527-1, January 27, 2006).

damage control on a flightline and other safety concerns. Currently, there is little to no reference to data integrity or cybersecurity in logistics issuances.[3]

Lower-level leadership (generally company-grade officers and senior noncommissioned officers) should develop clear, specific guidance to workers for what kinds of data anomalies they need brought to their attention based on their knowledge of operational needs.

All messaging should extend to all workers, to include airmen, civilians, and contractors. This messaging should persist over time. For airmen, it should start with technical schools and continue thereafter during on-the-job training and exercises. In addition, it should be performed, as much as is practical, in the work environment with real examples, and it should emphasize possible consequences of failure to detect and report suspected anomalous data.

Reporting and Evaluating

There are numerous inadequacies in the current reporting and evaluation mechanisms in the U.S. Air Force. None were designed specifically for mitigating cyber attacks. We recommend supplementing the current reporting mechanisms and assessments with some additional reporting mechanisms and processes and creating a coordination cell. The overall proposed structure is shown in Figure S.1.

In Figure S.1, existing structural elements (reporting channels and organizations) are depicted in black and blue and newly proposed structural elements in red. Reading the figure from the bottom up, airmen, civilians, and contractors (denoted by the box labeled "Reporters") report suspected incidents of corrupted data. The many existing reporting channels (e.g., help desks, Form 22 reports, 107 requests) are shown by the black line. These reports go to various organizations for some form of assessment (e.g., program offices, call centers, the Defense Logistics Agency) and often are routed through or reported to other organizations for situational awareness (e.g., major command staffs), as shown in the box to the right of the "Situational awareness" bracket. After assessment, some feedback generally flows back to the reporters. In a few cases, the feedback might result in the change of policy, education, or training, but remedies of that kind are not currently systematically performed.

---

[3] For example, the foundation policy for aircraft maintenance, Air Force Instruction 21-101, 2015, does not convey these messages.

Figure S.1. Proposed Response Structure for Reporting and Evaluating

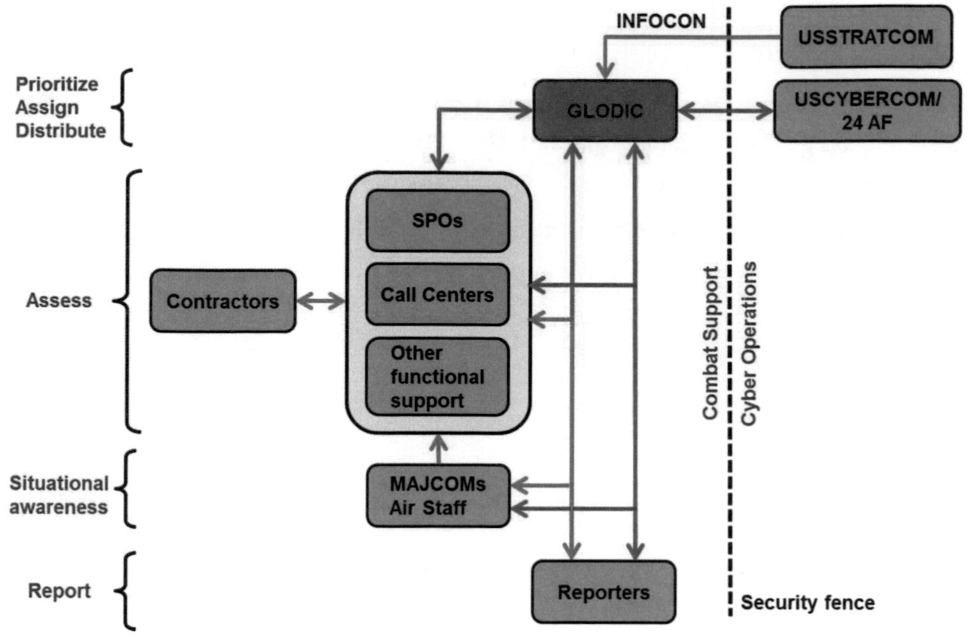

NOTES: Lines in blue indicate existing channels. Lines and boxes in red indicate proposed new channels and entities. GLODIC = Global Data Integrity Cell; MAJCOM = major command; SPO = system program office; USCYBERCOM = U.S. Cyber Command; USSTRATCOM = U.S. Strategic Command.

No authority has visibility of all of these currently performed reports in order to assess whether a coordinated attack might be taking place. Nor is there a central authority accountable for determining the cause of any significant corruption of critical data and disseminating causal information for remediation.

Hence, we propose a new central body that we call a *Global Data Integrity Cell* (GLODIC) that would receive copies of all such reports. This cell would maintain enterprise-wide situational awareness of all existing reports. Because these reports are being acted upon by the assessors, defined in the policies governing the current reporting mechanisms, most of the role of the GLODIC for these reports would be to look for trends and potential coordinated attacks, extract lessons learned over time, and serve as a clearinghouse of these lessons learned. This cell need not be restricted to this activity and should track other cyber-related logistics matters. It could be an existing organization that augments its responsibilities to include these, with a commensurate increase in resources.

The second proposed supporting activity is to assess whether a cyber attack has occurred. The assessment should be performed according to where the requisite expertise lies and does not have to be conducted by a central organization. For weapon systems, the locus of expertise might lie in the Air Force Materiel Command. For networked information technology systems, the locus of expertise might lie in 24th Air Force. But, as with the GLODIC, some organization needs to form an enterprise view, archive incidents, and disseminate instructions. Instructions might include whether to isolate systems from networks, change procedures, and lock systems for forensics. These functions could be performed by U.S. Cyber Command and/or 24th Air Force.

Prioritizing

Solving these issues across all systems and the functions that they support would be a large burden. For optimal response, proactive mitigations and specific reported incidents need to be prioritized. Incident response needs to be assigned a time frame for assessment and potential mitigation. The purpose of the prioritization is to rank criticality, not to rate it. A key element for prioritization is an understanding of the trade between any inherent robustness of a process (the time it takes for an attack to affect combat operations) and the resiliency of the response (the time it takes to detect, report, evaluate, mitigate, and recover) after an attack. Robust processes are a lower priority for response; fragile (not robust) processes are a higher priority. Time, viewed in this competing frame, and a proxy for the likelihood of an attack form two important dimensions of criticality that serve as a basis for our proposed prioritization. A strategy for prioritization would then unfold in this order:

1. List the (probably several dozen) functions performed within logistics. The level of indenture of this breakdown of functions will depend on the discovery during the following steps.
2. For each function, determine the principal data systems that control process or archive authoritative data on which actions or decisions are based.
3. Rank the susceptibility of these data systems by a proxy of the relative security of their hosts.
4. If erroneous data were acted upon in each of these data systems, estimate, roughly, the time it would take for a negative impact to combat operations (hours, days, weeks, months, or years).
5. If erroneous data were acted upon in each of these data systems, estimate, roughly, the response time, which might be dominated by the time it would take to recover (hours, days, weeks, months, or years).
6. Compare these times and rank the functions accordingly.
7. Plot the ranking of the functions' buffer time against the associated data systems' susceptibility.
8. Prioritize by working from the pairings of functions with the shortest buffer time and highest data-system susceptibility to the pairings of functions with the longest buffer time and lowest data-system susceptibility.

Of these recommendations, we assess that the two that are most promising are the introduction into logistics policy of responses to the various INFOCON status levels and the assignment of the responsibilities we outline to an organization we call the GLODIC. This organization can be an existing organization and can also perform other cyber-related functions, but it should be able to handle data corruption.

Because data corruption is just one type of cyber attack, we developed the main arguments in this report as much as possible to be sufficiently general to address the full spectrum of cyber attack. The prioritization scheme can be easily extended to the full spectrum of cyber attack. The ideas for enhancing detection are not as critical for some cyber attack types, such as destroying data and denying access, because these attacks are quite obvious, but they would be useful nonetheless for early detection of less obvious attacks. The methods we recommend for encouraging reporting and evaluating those reports would work for all cyber attack types and should not be restricted solely to data corruption.

Acknowledgments

We thank Larry Kingsley and Lorna Estep for sponsoring this work and for their support throughout. Many others in the Air Force helped and supported us, too numerous to mention by name. Their ideas and critiques were indispensable.

At RAND, we thank Pat Boren, Lionel Galway, Lauren Kendrick, Sarah Nowak, Marek Posard, Jim Powers, Marc Robbins, and Lara Schmidt for helpful discussions.

Dan Gonzales and Danielle Tarraf provided useful reviews of an earlier draft.

That we received help and insights from those acknowledged above should not be taken to imply that they concur with the views expressed in this report. We alone are responsible for the content, including any errors or oversights.

Abbreviations

AFMC/EN	Air Force Materiel Command Directorate of Engineering and Technical Management
ASRS	Aviation Safety Reporting System
CYBERCON	cyber conditions
FPCON	force protection condition
GLODIC	Global Data Integrity Cell
INFOCON	information operations condition
LogATSO	Logistics Ability to Survive and Operate
MAJCOM	major command
MICAP	mission-capable
NIPRNet	Nonsecure Internet Protocol Router Network
PAF	Project AIR FORCE
PKI	public key infrastructure
SPO	system program office
USCYBERCOM	U.S. Cyber Command
USSTRATCOM	U.S. Strategic Command

1. Approaching the Problem

Motivation

The U.S. Air Force can, within a short time, project formidable air power nearly anywhere in the world. This air power poses significant challenges for an adversary to attack directly. But air power requires timely logistics support,[4] and therefore an adversary might choose to attack air power indirectly by targeting its logistics support. The goal would be to cripple U.S. air power by undermining its foundations. The desired effects could range from slowing or impeding the ability of air power to establish expeditionary operations—perhaps just long enough for the adversary to achieve its desired ends—to ceasing all combat operations by paralyzing logistics.

To safeguard the ability to project air power, the U.S. Air Force must ensure that its logistics capabilities are able to absorb an attack (robustness) and recover functionality satisfactorily afterward (resiliency).[5] Attacks to logistics can come in many forms. Traditional forms of attack have been kinetic—one salient example is the bombing of runways. Although the logistics community is not responsible for directly protecting runways from bombing, it is responsible for rapid runway repair capabilities to recover operations. There are many such examples, and the U.S. Air Force has numerous capabilities to ensure an acceptable level of logistics support despite kinetic (including chemical, biological, and nuclear) attacks.

Cyberspace presents a relatively new domain for attack. As with most kinetic attacks, the logistics community is not directly responsible for protecting its systems from these attacks, but it is responsible for being able to work through such attacks and recover afterward.[6] Cyber operations can take many forms. An adversary might exfiltrate information for intelligence purposes, or an adversary might attack offensively by destroying data, corrupting data, denying access to data, or taking control of processes. All these forms of cyber operations present risks to logistics.

Logistics in the U.S. Air Force is performed by a range of functional communities, including aircraft maintenance, airfield operations, civil engineering, logistics planning, munitions management, security forces, and supply chain operations. To do these functions, hundreds of information technology systems have been developed to store and disseminate data, to manage

---

[4] We use the term *logistics* throughout this report in the broadest sense, as defined in Joint Publication 1-02, 2016: "Planning and executing the movement and support of forces." The concepts are general enough that they apply equally to *logistics* and *combat support*; Air Force Doctrine Document 4-0, 2013, defines the latter as "the foundational and crosscutting capability to field, base, protect, support, and sustain Air Force forces across the range of military operations."

[5] See, for example, Defense Science Board, 2015.

[6] There is a parallel with the discipline of safety, in which awareness, diligence, and responsiveness by every single person is essential. We will discuss these attributes in Chapters Two and Three.

stock inventory, to control processes, and to provide situational awareness to support decisionmaking. These information systems are used continuously worldwide at U.S. Air Force installations in both peacetime and wartime. Many of these systems are interconnected, and although most are unique to the U.S. Air Force, some connections extend to the joint community and to industrial base partners. Nearly all reside on the Nonsecure Internet Protocol Router Network (NIPRNet).

Logistics operations rely heavily on the data and information in these systems.[7] In the absence of reliable data, many key functions would be halted. The sheer size and complexity of the supply chain place demands on knowledge of the identity of parts, stock levels, and part locations—and many other data—that exceed human capacity. These operations require data systems to function; without these data, aircraft operations would eventually come to a standstill. Another example is aircraft maintenance. Modern aircraft are too complex to be maintained without reliance on technical data and automated diagnostic equipment.

Of course, data errors are inevitable. Errors occur routinely from everyday mistakes. For the most part, these day-to-day errors do not have significant negative operational impacts. Logistics systems and processes have evolved to handle them. And the randomness of routine errors makes it unlikely that any one error will cascade into a major operational problem. But significant impacts are possible, as experience has shown. Two recent examples are the sending of a valid part for an operational weapon system to be demilitarized and the execution of an incorrectly specified time compliance technical order on a fleet of aircraft.

A skilled, determined, and knowledgeable adversary could potentially wreak far more damage by deliberately corrupting data that are unlikely to be detected as anomalous, yet targeting the attack to have a significant negative impact on operations. An adversary might choose this kind of targeted attack by corruption over data destruction or denial of access to data in order to maintain a longer foothold in the systems or to mask attribution. Regardless of whether data are corrupted by attack or random error, logistics support should be sufficiently resilient and robust to data corruption to continue providing adequate support to combat operations.

Purpose and Scope of Report

The purpose of this report is to examine ways to improve the continuity of logistics operations when logistics data have been corrupted. By corruption, we mean an alteration of data that is significant enough to have a negative operational impact. Potential examples include an erroneous shipping destination for critical spares, incorrect technical data for procedures for aircraft repair, a false diagnosis from automated test equipment, a spurious redistribution order

---

[7] Frequently, the term *data* is used to describe raw numbers and codes in databases, and *information* is used to describe meaning extracted from data to make decisions. For simplicity, this report uses the term *data* in its generic sense throughout, which also includes information.

for spare parts, and countless others. In this report, we do not assess which of these adversarial cyber operations are of highest concern to logistics.

The cybersecurity community provides a first line of defense against these threats. However, should those measures fail, the logistics community is responsible for operating through and recovering from attacks without adversely affecting combat operations. The scope of this research was further limited to actions that could reasonably be executed by the *logistics community* (and not those that would be executed by the cybersecurity community) to provide resiliency to operations. A final constraint was to avoid solutions that are costly and would likely take years to implement. Therefore, we restricted the discussion and potential solutions to those that are economical, relatively easy to implement, and within the scope of the logistics community.

Before outlining our approach to the problem of data corruption, we state explicitly a few assumptions and premises. We take as givens that *all* data are prone to corruption and that *all* data systems are susceptible to attack. This susceptibility extends beyond business and networked information technology systems to include weapon-system support systems and weapon systems themselves. Many information technology systems that support logistics reside on relatively unsecure networks and exchange data with numerous commercial entities whose security postures are not controlled by the U.S. Air Force. But even systems that are nominally unconnected to outside networks exchange data externally, and they, too, are susceptible to attack.

Because these systems largely reside on relatively unsecure networks, we do not attempt to assess the overall vulnerability of these systems to cyber attack. But for the purposes of developing robust and resilient logistics operations, all systems must be considered to be at risk. Just as no base is utterly secure against kinetic attack, no system is utterly secure to cyber attack. We will not argue this point further because experience has indicated that even systems thought to be highly secure have been compromised. Nevertheless, in the course of this study, we encountered numerous individuals within the Air Force logistics enterprise who believe that the data systems that they use are nearly invulnerable to attack. Such overconfidence might prove to be a barrier to improving the robustness and resiliency of all operations and will be discussed in later chapters.

We also take as a premise that the ultimate goal of this endeavor is the mission assurance of combat operations (e.g., sortie generation). The continuity of logistics operations is important to the degree that logistics operations affect combat operations. A final premise is that the number of functional areas and data systems susceptible to corruption is too large to address immediately. Limited resources mean that some prioritization is needed to balance the factors that affect robustness and resiliency, a topic we take up in Chapter Five.[8]

---

[8] See Snyder et al., 2015.

Approach

Context

Data and the computer systems that provide situational awareness and control processes greatly enhance logistics capabilities. Many large, complex operations have evolved to depend on data and cannot be adequately controlled by humans alone. Operators must either trust data or rely on their limited knowledge and capabilities and risk making mistakes. But data and information can contain errors, and algorithms that provide situational awareness and control processes can have bugs or fail. In these cases, operators who blindly follow data are also at risk of making a mistake. The trust that operators place in technology must be well calibrated with the overall accuracy and reliability of the data.

Figure 1.1 depicts this ideal balance between trust and suspicion in data. Along the x axis is plotted the accuracy of data (reality) and along the y axis the trust workers place in that data (attitude of the workforce). The blue lines represent well-calibrated trust in the data—little trust for data of low accuracy and high trust for accurate data. One line indicates well-calibrated trust during routine peacetime operations, and the other indicates well-calibrated trust during wartime, when the threat of data corruption is higher. Workers with excess trust fall in the upper left region of the plot. These workers are at risk of making mistakes by blindly following data that are in error. One example of this first kind of fault is making a sequence of dangerous flight maneuvers because a pilot incorrectly trusted a malfunctioning airspeed indicator and incorrectly deduced that his plane was stalling when it was in fact flying normally.[9] Workers with excess suspicion fall in the lower right region of the plot. These workers are at risk of making mistakes by substituting their limited judgment and skills for more accurate data and potentially slowing operations. One example of this second kind of fault is the midair collision of two aircraft when the pilots of one aircraft erroneously relied on human judgment over a correctly operating automated collision avoidance system.[10]

---

[9] Croft, 2016.

[10] Bennett, 2004.

4

Figure 1.1. Schematic Diagram of the Balance Between Trust and Suspicion

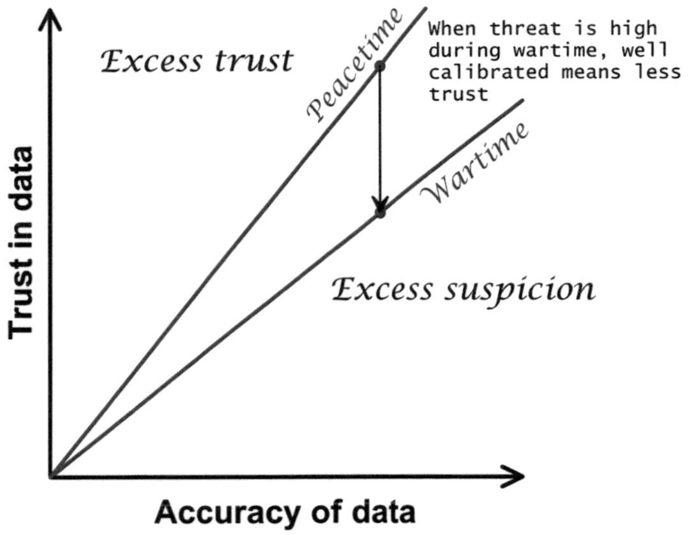

Figure showing two axes: "Trust in data" (vertical) and "Accuracy of data" (horizontal). Two diagonal lines labeled "Peacetime" and "Wartime" emanate from the origin. The region "Excess trust" is labeled above, and "Excess suspicion" below. An annotation reads: "When threat is high during wartime, well calibrated means less trust."

NOTE: This perspective is inspired by Lee and See, 2004, p. 55, Figure 2.

Because the degree of trust for a well-calibrated response depends on the accuracy of the data, it not only varies from system to system but also varies for a given system over time. It is expected that the threat to the integrity of data in a system will be greater in times of war (and leading up to war) than in times of peace. As the threat of attack increases during the lead-up to war, workers need to recalibrate their trust to adjust to the potential for corrupted data. This recalibration might temporarily place workers in the "excess suspicion" field of Figure 1.1 as they anticipate possible data corruption in a transition to a wartime footing. Care will be needed to avoid the self-infliction of mistakes caused by excess suspicion of data.

For workers' trust to be well calibrated, they must understand both axes of the plot in Figure 1.1. That is, they must understand the data they handle well enough to assess the accuracy of the data, and they must understand their own knowledge and its limits.[11] To understand the accuracy of the data, they must first acknowledge the possibility that data could be corrupted beyond the day-to-day errors that they normally experience. They must understand that the data *can* be attacked. If they are in denial of the possibility of errors in data, they are at risk of drifting into the "excess trust" field in Figure 1.1, where they might not detect a significant error when it occurs, before the ensuing negative impact.

But workers must also have knowledge of the processes they use when evaluating data. They need to understand what they know and the bounds of their expertise. If they believe that they know more than they really do, they risk falling into the "excess suspicion" field of Figure 1.1 and making a mistake or overreporting suspected corrupted data. If they underestimate their

---

[11] See Lee and Moray, 1992.

knowledge, they risk falling in the "excess trust" field and either making a mistake or underreporting suspected data corruption.

In Chapter Three, we discuss in more detail ways to keep workers well calibrated with regard to trust in data.

Framework

An effective response to a disturbance in a process requires accurate feedback to actors who make decisions and act to adjust the state of the process. In 2015, as part of an initiative called Logistics Ability to Survive and Operate (LogATSO), Air Force logistics leadership described three steps for response: (1) detect potential errors; (2) evaluate whether these are indeed errors, and, if so, determine whether they are merely mistakes or the result of a deliberate attack; and (3) mitigate any problems to logistics operations created by the real errors. This decomposition captures the feedback and decision-loop characteristics of a dynamic response, and we adopt this framework.

This report fleshes out the LogATSO framework for responding to data corruption. The goals of this effort are (1) to increase the robustness of combat operations by increasing the robustness of logistics operations in the face of corrupted data, (2) to improve the overall cybersecurity of the U.S. Air Force by diminishing the impacts of an attack on logistics, and (3) to deter an adversary from embarking on such an attack by having a credible response and projecting that capability. The emphasis in this report will be on the overall structure for how the U.S. Air Force can maintain robust and resilient logistics operations in the face of data corruption, where it should prioritize its efforts in this area, how anomalous data can be better detected, and how to conduct evaluations to formulate an appropriate mitigation at the enterprise level. Although identifying potential mitigation measures is out of scope for this report, we offer some measures that emerged during the research into detection and evaluation of data corruption.

In the next three chapters, we discuss detection, reporting, and evaluation in some detail. We then tackle the issue of where to focus effort in a resource-constrained environment in Chapter Five. Where are the data issues of most concern, and, therefore, where should efforts be directed first? Where can risk be accepted? This discussion on prioritization is followed in Chapter Six by a brief but broader discussion of the cyber threat and how the logistics community can be better postured to be robust and resilient.

2. The Challenges of Detection

Detecting an error is a necessary first step before taking any mitigating actions. To improve detection of anomalous data, we need to understand clearly at what juncture detection needs to have happened to avoid negative consequences and what actions need to occur for detection to be successful. In this chapter, we analyze these topics and conclude with a list of key areas to improve detection of corrupted data. In Chapter Three, we then offer recommendations for improving detection in these key areas.

The Promptness of Detection

Before any evaluation or corrective action can be taken in response to corrupted data, the anomalous data must be detected. It is safe to say that if corrupted data will cause some operational harm, they will eventually be detected. But one of the key observations from the last chapter is that if detection does not happen until operations, such as sortie generation, are halted, the detection will have come too late. Therefore, it is important to detect corrupted data that could have a negative combat operational impact early enough to respond before the negative impact to combat operations can occur.

Two avenues present opportunities to improve promptness of detection: (1) quicker detection by an individual, such as by enhanced attentiveness, and (2) earlier detection in a chain of custody of data, such as detecting an anomaly in technical data by a program office before the maintainer even sees it and could detect it. The earlier in a chain of custody that an anomaly can be detected, the less reliance on individuals is needed to detect it quickly.

The Elements of Detection

To analyze detection, a useful starting point is to define what we mean by *detection*—what has to happen for detection to occur. For an erroneous datum to be detected, several conditions must be met. First, someone or something that can act (an *agent*) must be exposed to the erroneous data. Second, the error must be an anomaly, which means that it must differ from some expected baseline. Third, the agent must have sufficient expertise to distinguish an anomaly from the expected baseline pattern. And fourth, if humans are to be detectors (sensors), they must have some motivation to detect and report an anomaly—or, at the very least, they must not have any incentives to hide the anomaly from reporting. We discuss these conditions in the following sections.

Exposure to an Agent

If no agent sees the anomalous data, no detection can occur. This exposure can come from seeing the data during the agent's normal activities (reactive detection) or through active probing by the agent looking for anomalies (proactive detection). Agents can be humans (airmen, civilians, and contractors) or computers (algorithms). We focus on humans as detectors to stay within the bounds of the research question we were asked to address. In the next chapter, we discuss one way in which an algorithm can assist in detection, but only in the context of a tool for humans to use as an aid in their detection skills.

Human exposure to data varies in some important ways in this context. In some cases, humans regularly see data. Many times, the point of the data is to direct human actions. An example is the use of technical data to instruct maintainers in aircraft maintenance procedures. For experienced maintainers who are following technical data that they have followed numerous times in the past, it is reasonable to expect that anomalies might be detected. If a maintainer normally tightens a bolt to a certain torque, a journeyman should have enough experience to detect the anomaly if there is a sudden, unannounced change in that torque by a factor of two. But a novice performing the task for the first time would probably not detect it—the novice would not recognize the anomaly because the novice would not know the expected baseline.

In other cases, a human could be exposed to anomalous data but cannot reasonably be expected to detect that anomaly because the person does not have enough contextual knowledge, regardless of experience level. For example, if an aircraft part whose type is currently in use is noted (in error) in a supply database to be demilitarized, the worker executing the task of demilitarizing the part cannot be expected to know that that part should be held in supply, not demilitarized. That level of knowledge exists only at higher levels in the organization, and detection of this kind of anomaly can only be expected to occur where that knowledge resides. In some cases, the requisite knowledge to detect an error might lie outside the Air Force (e.g., in the defense industrial base).

The situations in which workers lack the knowledge to be reasonably expected to recognize an anomaly are increasing and are unlikely to decrease in the future. Systems, such as the F-35A, are highly complex, and both supply and maintenance personnel are reliant on information technology to diagnose, manage, and execute their processes. This reliance is a natural outgrowth of complexity of the systems themselves and the capabilities of modern information technology. With these changes have come decreases in manpower levels and less direct experience with the details of the data. A modern mechanic uses a computer to diagnose malfunctions and no longer relies solely on experience, such as listening to the sound of an engine or tactile feedback. Because of this reliance, the skills for diagnosis without a computer have atrophied in the workforce. So, too, has a similar phenomenon followed the technological changes in logistics operations.

In yet other cases, data are not seen at all by humans during normal operations. Not being exposed to any anomalies, humans cannot detect them. Many parts of the operation of the supply chain are automated to the extent that many transactions are never monitored by a human.

To have robust and resilient logistics operations in the face of data corruption, effective detection mechanisms are needed for each of these three cases. The most challenging case for exposure to an agent is the one in which no human sees the data in normal operations. The case of workers seeing data but not having sufficient expertise is taken up later in this chapter in the section on expertise. The case of workers seeing the data and having the expertise presents challenges for motivation and is also discussed in that section.

Anomalies in Data

If corrupted data are not noticeably different from some baseline—if they are not anomalous—humans cannot detect them. In this context, three factors can hide anomalous data and render corrupted data difficult to detect.

First, some data sets have high variance, and, therefore, it takes a large deviation from some expected baseline for anomalous data to be noticed. In other words, there is so much noise in the normal data that seeing something odd is difficult. Perhaps the most salient example of a high-variance data environment is the supply chain. For many parts, the frequency of demand is episodic, and their shipping destinations are variable. In addition, the volume of supply chain operations is huge. Therefore, the baseline of normal operations has many fluctuations, and the number of data that provide a background against which to observe anomalies is vast. These attributes obscure anomalies, and an adversary might exploit this situation to hide a data-corruption attack.

On the positive side, because of the intrinsic high variance of the data, supply chain operations have evolved to be robust against a certain level of errors. Small errors are in fact day-to-day events in supply chain operations. Hence, a small anomaly, although not easily detected, also generally causes only a small impact to combat operations. The normal operations of the supply chain can generally absorb small errors without undue operational impact.

However, some small errors can have large effects, which leads to a second factor: Data corrupted by an adversary during a cyber attack are likely to present anomalies that will differ from the errors that workers typically see. Targeted attacks of concern would be designed to have as many negative effects on operations as possible with the smallest changes. A targeted attack pattern would probably differ from typical, day-to-day errors in a data system, which tend to be of random origin. It is also nearly impossible to predict in advance what corrupted data patterns from an attack might look like because of the enormous number of possible data to be corrupted. It is likely that data anomalies will differ from those of prior experience or analysis. To the extent that the baseline changes during wartime, the difficulty of detection increases.

The third obscuring factor arises because the baseline that workers have for discerning anomalous data is the routine activity of steady-state operations (e.g., during peacetime). This

9

baseline is not the same as that of high operations tempo (e.g., operations during peak time-phase deployment activities or peak sortie generation). Yet this second, wartime scenario is potentially the most dangerous. Attacks during peacetime might be a nuisance, but they are unlikely to have grave impact on combat operations. In fact, if attacks during peacetime had a serious impact on combat operations, they might very well elicit transition to a wartime posture. Attention to cyber attacks is therefore most acute during wartime and its precursors.

Sending critical F-22 spare parts, for example, to a base in which F-22s do not operate could attract attention during peacetime. But during expeditionary deployment, that same shipment might not be challenged because workers might very well assume that F-22s were deploying to that location and they simply had not been informed. Activities during expeditionary deployment are by nature unlike those that workers will have previously experienced, even if they are veterans of previous expeditionary operations.

These three factors that can hide corrupted data by diminishing anomalous characteristics all present challenges for detection. Because the largest concern for cyber attacks is during the various phases of wartime, we judge the third obscuring factor, that of detecting wartime anomalies when the experience baseline is acquired in peacetime, to present the most severe challenge. Said another way, the most challenging problem presented by the need to detect anomalous behavior is to identify what constitutes anomalous behavior, and during the most critical time—wartime operations—the baseline itself is anomalous.

Expertise of the Workforce

Even when humans are exposed to anomalous data, they need to have enough expertise to understand what an expected range of the data is and whether the observed degree of out-of-range behavior is enough to have a significant, negative effect on operations. They need to have the skills to recognize the anomaly and to judge whether it needs to be reported to a higher level (supervisor, call center, etc.). The less obviously anomalous the corrupted data, as discussed in the previous section, the more expertise is required to detect it. Expertise is acquired through training, education, and experience, but it can only be effective if workers continuously monitor data, learn normal patterns, and associate real-life consequences with the corresponding erroneous data. They need to be *mindful*, by which we mean that they need to scrutinize ongoing operations, continually refine their understanding of operations, and possess a willingness to challenge their own expectations.[12] The challenge for expertise, then, is to develop and sustain a culture of mindfulness—one that spans the full work environment, from basic training, to the technical schools, and on the job. This culture must also extend beyond airmen to include civilians and contractors.

---

[12] See Weick and Sutcliffe, 2001, p. 42.

Motivation to Report

Workers who are exposed to corrupt data that is recognizably anomalous and who possess the expertise to detect it have one further barrier: They need to be motivated to report it—or, at the very least, they need to have no disincentives to reporting. Some natural human tendencies work against the desired motivation to report anomalous data. Some workers will feel that reporting suspected anomalous data is akin to pointing out potential problems in their operations. They might not want to reveal problems to their supervisors, preferring to project an image that they have everything under control. This tendency can lead to waiting to report a detected anomaly until the problem has reached a crisis point beyond the control of the worker when it is too late to mitigate the attack. A near miss will then have become an accident. This phenomenon can also manifest itself in organizational units. Units can want to internalize their problems and can be reluctant to report issues externally for the same reasons.

This tendency will almost surely be amplified just at the most critical time—during wartime operations. During wartime, leaders need and expect workers to get the job done and not complain about issues they are facing. Leaders are under their own pressures during these moments, and the workers under them know that. They might fear that questions about data might be poorly received by leaders. When, exactly, it is appropriate to report? Will superiors become irritated if perceived anomalies are reported and then turn out to be okay? Will the worker be seen as someone not getting the job done but instead delaying operations by questioning too much? How can workers feel free to question the data and information they have been given?

This phenomenon can be especially problematic for a coordinated attack against a weapon system. Suppose that a weapon system is attacked by corrupting weapon system maintenance data, the supply system, the transportation of those supply parts, and the base operating support where the weapon system is located. Each of these attacks might look small to each of the stovepiped functional areas and, indeed, might not even be recognized as an attack, just as a problematic anomaly. But to form an enterprise view of the situation and evaluate it as a coordinated attack, workers in each of these functional areas need to be motivated to report out their observations so that someone with a broader scope of vision can assess the situation.

As systems become more complex, data become larger, and more tasks become automated, workers are being trained to increasingly trust data. But the task here is to also view data with some level of suspicion. This conflicting view of trust and potential distrust of data is a tension that must be overcome by a higher mutual trust between the workers and their leadership.

Workers also need widely known, formal channels for reporting the data. These reporting channels are a critical part of the evaluation process. Because of this need to integrate reporting channels with evaluation mechanisms, we discuss reporting channels in Chapters Three and Four.

Solving motivational issues requires a cultural change. To enhance detection and reporting, a culture of trust needs to be cultivated and nurtured, one that gives airmen, civilians, and contractors confidence that their actions will benefit the mission and that no retribution will befall them for acting and reporting.

Summary of the Challenges of Detection

Five critical areas show promise to enhance the ability to detect corrupted data:

- The detection must be sufficiently prompt. Promptness results from a combination of individuals detecting and reporting corrupted data quickly and detecting corrupted data early in the chain of custody.
- Highly automated processes—in which humans do not see the data during normal operations—present a significant challenge for detection. Automation requires special mechanisms to assist in detecting corruption.
- Detection of corrupted data is most critical during wartime, yet anomalies are less evident during wartime than during peacetime because wartime itself is an anomaly. Mechanisms are needed to adjust detection mechanisms from peacetime to wartime conditions.
- For workers (airmen, civilians, and contractors) to detect anomalous data, they all need to be trained to understand the expected baseline and need to be continuously vigilant when examining data.
- Leadership must create an environment that encourages workers to report suspected anomalous data.

In the next chapter, we propose ways to at least partially address these critical challenges and encourage associated cultural shifts to facilitate anomaly detection.

3. Recommendations for Improving Detection

There are numerous actions that can be taken to address the challenges of detecting corruption in data. They range from relatively detailed recommendations for individuals at the working level to policies and guidance for leaders to change. We discuss these recommendations in a structure that parallels the previous chapter's discussion of challenges. It will become clear, however, that some of our individual recommendations at least partially address multiple challenges and that individual challenges are best addressed through multiple lines of action.

Improving Promptness

Promptness of detection can be accelerated at two levels: at the individual level by changing the behavior of workers and at the enterprise level by detecting errors earlier in the handling and passing of data rather than at the stage just before actions are taken based on the data.

Individual-Level Actions

Data exist to inform decisions, and decisions are made to effect actions. The risk of a data error, therefore, is the risk of an error leading to a poor decision, which, in turn, leads to an action counter to operational goals. It is critical for individuals to detect any significant anomalous data before actions are taken. Three factors stand out in accelerating an individual's time to detect anomalous data.

First, if a worker is expected to detect and report anomalous data early (before actions based on that data are taken), the worker needs to be exposed to the data early. If the worker only sees the data immediately before (or worse, after) actions are taken, it is likely to be too late to evaluate, mitigate, and avoid negative operational impact. Yet exposing workers routinely to data before they would see the data in their normal activities would be onerous and could intolerably slow down logistics operations. However, circumstances might very well exist during a heightened threat environment (e.g., during wartime) when certain high-priority areas (identified by the process outlined in Chapter Five) would benefit by deliberately instructing workers to examine some data earlier than they ordinarily do. We will further discuss this idea in the section on adjusting to wartime later in this chapter.

Second, quick detection and reporting would be facilitated by leaders creating an expectation in the workforce to detect and report suspected corrupt data. This cultural shift, much like the culture of reporting safety concerns, is a strong underlying theme necessary to address many of the challenges of detection and will be discussed more fully in the section on motivating detection later in this chapter.

The third factor is to make it easy to report suspected data corruption. A considerable literature upholds the observation that one of the most successful attributes of programs to get people to do something is to make it easy for them to do it.[13] Beyond exposing workers to data more often, there might not be many ways to make it significantly easier for a worker to detect anomalous data. However, the reporting of their suspicions can be made easier by providing easy and familiar reporting mechanisms. We take up ideas for reporting mechanisms in Chapter Four.

Enterprise-Level Actions

A lot of data move from one organization to another before they are used to inform a decision. Technical data for maintenance, for example, might originate in a program office, flow to a major command, and then move to the staffs of a wing and group before finding its way into the hands of a maintenance worker who will, based on the data, execute actions. We call this flow a *chain of custody*. Clearly, the earlier errors can be detected in the chain of custody, the shorter the response time and the more resilient the operation. Consistently identifying data errors early in the chain of custody requires the involvement of the entire enterprise, including airmen, civilians, and contractors. It is better to detect an error in technical data in a program office rather than on the flightline, and it is better that an error in shipping be detected in supply, if possible, rather than in transportation. Not only is the time available to respond decreased when it is left to the last person in the chain of custody to detect an error, but it will also often be the case that the last person does not possess all the requisite knowledge to discern the anomaly. Therefore, all the other recommendations made in this chapter benefit from the widest possible application throughout the enterprise, spanning all airmen, civilians, and contractors.

Overcoming Automation

Modern logistics operations are a complex partnership between humans and technology. Sometimes the technology takes a leading role in this partnership. The operation of the supply chain is one such area. Especially in the portion managed by the Defense Logistics Agency, the supply chain has, by design, many processes that are nearly fully automated, with little to no human participation. In these cases, precious few opportunities occur for a human to detect a data error before actions based on that data are executed. Situations like this one have become more common as logistics processes become more complex and additional efficiencies are sought. Automation requires means other than human sensors for error detection.

The idea explored in this section is another form of human detection, albeit with the assistance of a tool—methods for spotting certain kinds of data corruption. The methods for detecting anomalous data are numerous. We limit the discussion to techniques that are simple and not costly to implement. All the methods that we describe exploit that fact that useful data

---

[13] Thaler and Sunstein, 2008.

have patterns. These patterns are not easy to reproduce when data are corrupted, making it somewhat difficult to fake data.[14]

Benford's Law

One method that is commonly used in fraud detection is based on a common pattern in data called *Benford's Law*.[15] The leading significant digits[16] in a surprisingly large number of data sets are not uniformly distributed. The digit 1 occurs more frequently than 2, and 2 more frequently than 3, and so on, such that 9 is the least frequently occurring leading significant digit. If we call the first significant digit D and d is any integer in the set $\{1, 2, 3, \ldots, 9\}$, Benford's Law for the first significant digit is[17]

$$\text{Probability}(D = d) = \log_{10}\left(1 + \frac{1}{d}\right).$$

Data sets following this distribution are said to be *Benford*. If a data set is Benford, then by this equation there is approximately a 30-percent chance that the first significant digit of any datum in that data set is 1, about an 18-percent chance that the first significant digit is 2, and so on, decreasing to only 4.5 percent for the first significant digit being 9. When altered fraudulently, Benford data sets depart from this pattern, a fact that is used in fraud detection.[18]

An astonishingly large range of data types are Benford. These vary as widely as physical constants, radioactive decay rates, Internal Revenue Service files, certain stock market data, and many more.[19] When should we expect a data set to be Benford? There is as yet no clear answer. However, it can be proven mathematically that taking any distribution and repeatedly multiplying or dividing by random numbers or raising it to a random integral power numerous times converges to a Benford distribution. And a distribution that is Benford remains Benford under multiplication, division, and the raising to a power (scale and base invariance).[20] In fact, a distribution that is scale invariant is always Benford. This invariance means that if a distribution is Benford when expressed in one set of units, it is Benford when expressed in any units.

Researchers have laid out some properties that tend to lead to a data set being Benford, though satisfying these conditions will not guarantee it:[21]

---

[14] See, for example, Hill, 1999.

[15] Benford's Law is actually a commonly observed pattern, not a law.

[16] The *first significant digit* in a number is the first nonzero digit when reading from left to right. For example, the first significant digit of 218.81 is 2 and that of 0.0375 is 3. The first significant digit is always nonzero, but second and higher significant digits can be 0. The second significant digit of 0.102 is 0.

[17] If we let D_n be the nth significant figure reading from left to right, then the general form of Benford's Law is $\text{Probability}(D_1 = d_1, D_2 = d_2, \ldots, D_m = d_m) = \log_{10}\left(1 + \left(\sum_{j=1}^{m} 10^{m-j} d_j\right)^{-1}\right).$

[18] Durtschi, Hillison, and Pacini, 2004; Nigrini, 2012; Berger and Hill, 2015, Chapter 10.

[19] Berger and Hill, 2015, pp. 4–5.

[20] Boyle, 1994; Hill, 1995; Berger and Hill, 2015.

[21] See Durtschi, Hillison, and Pacini, 2004.

- numbers coming from mathematical combinations of other numbers (e.g., products of numbers, such as price times quantity)
- transaction-level data (as opposed to aggregated data)
- large data sets that span multiple orders of magnitude in values
- data for which the mean is greater than the median and skewness is positive (long right tail)
- scale invariance.

Data sets that are less likely to be Benford are those composed of assigned or sequential numbers (e.g., telephone numbers), data that are influenced by psychological factors (e.g., prices set at $19.99), data with a large number of firm-specific numbers (accounts set up to record refunds of a fixed price, etc.), or data with a built-in minimum or maximum. Data that are presented as percentages rather than raw values are also less likely to be Benford. Data that have a fixed number of digits for each entry (for example, nine-digit National Item Identification Numbers) are often not Benford.

In a search for data following Benford's Law, we did an informal check of a range of logistics data and found some that were Benford. One example we found was pricing data from the Secondary Item Requirements System, often referred to by its data system designator, D200A. Figure 3.1 shows those data in comparison with an ideal Benford distribution. They are visually very close to being Benford. Numerous goodness-of-fit and other statistical tests confirm that the data are Benford.

Figure 3.1. Pricing Data from D200A That Follow Benford's Law

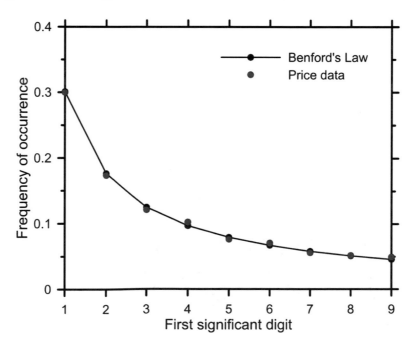

However, a lot of data that we examined did not conform to Benford's Law, including other data from D200A. One example is a D200A data set for reparable spare parts demand for aircraft for fiscal years 2008–2013. These data satisfy most of the conditions for Benford suitability in the section above, such as the size of the data set (nearly 118,000 records), spanning multiple orders of magnitude, and having a mean larger than the median with positive skewness. However, upon closer inspection, it fails some rigorous statistical tests. One test that is especially revealing is that it is not scale invariant. Figure 3.2 shows the original D200A data (raw data), the data multiplied by 2 (scaled data), and Benford's Law.[22] As can be clearly seen, the distribution changes significantly after multiplication by 2 and is therefore not scale invariant, and, therefore, it is not Benford.

Figure 3.2. Reparable Spares Data from D200A That Do Not Follow Benford's Law

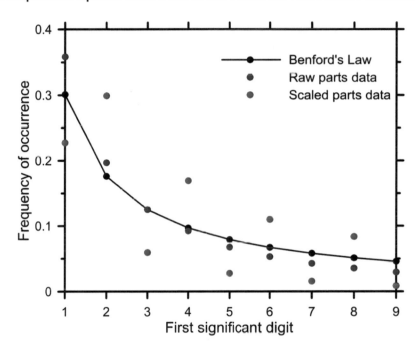

In tests with simulated data that conform to Benford's Law, and in which random points were then automatically modified to be non-Benford, we found that we were able to identify when approximately 9 percent or more of the data were changed using various goodness-of-fit tests. Hence, we expect that if an adversary altered more than 9 percent of the data in a Benford data set, statistical techniques could detect that level of corruption.[23] On the other hand, corruption of less than 9 percent of the data would be difficult or impossible to detect by this method.

---

[22] See Berger and Hill, 2015, p. 73, for how to apply this scaling test.

[23] See Nigrini, 2012, for a good review of tests.

Other Techniques

Even if a data set is not Benford, it often will nevertheless display a consistent pattern over time of the distribution of significant digits. If this pattern is known from sampling over time prior to any tampering, an analyst can compare suspect data to this known pattern. This approach would be very similar to the Benford's Law method described above, only comparing the data to a known distribution other than the Benford distribution. The advantage of Benford's Law is that prior researchers have established statistical techniques for detecting data corruption. Assessing whether a Benford data set has been significantly altered, therefore, can be made routine using established tools and does not require any special skills for the analyst. If the data set is not Benford, depending on the nature of the patterns of significant digits, ad hoc statistical methods might be needed and would therefore require the analyst to have advanced skills and develop tests specific to each data set.

We note one additional statistical test, *statistical process control*. Rather than examining a data set as a whole for corruption, this method examines each datum in a data set as a possible anomaly. The method is quite useful for time sequences of data, such as requisitions for a part. The method measures departures over time from any stable distribution of values for a data type. The general idea is to keep track over time of a certain statistic (a running average for number of requisitions, for example), as well as any variation in that statistic (such as the standard deviation or absolute deviation). From these, the method creates a control chart to illustrate whether any data are, in some sense, "out of bounds." The control chart plots three lines as a function of time: the center line (usually the average), an upper control limit, and a lower control limit. The upper and lower control limits are commonly three standard deviations above and below the mean. When the data do not follow a Gaussian distribution, other criteria are often used for these limits, such as the 95th and 5th percentile values. The limits provide a way to detect any outliers or exceptions.

Figure 3.3 shows an example of this method for Ogden Air Logistics Complex requisitions of O-rings (NIIN 013313891) ordered in the 65-month span from January 2011 through May 2016 recorded in the Strategic Distribution Database. The actual requisitions are shown in black. The solid blue line shows the running average from the previous 12 months (except for January 2011 through December 2011, which is the running average from the available previous months of data starting from January 2011). The broken red lines are the running three standard deviations obtained from the same data. Red dots indicate outliers that fall outside of the three standard deviations. There is an outlier early, at five months, which can probably be ignored because so few data entries are used to compute the mean and standard deviation for that time point. There are three outliers at months 42, 45, and 52, corresponding to requisitions in June 2014, September 2014, and April 2015, in which the numbers of requisitions were 23, 19, and 24, respectively. These outliers are excluded in later computations for running average and standard deviation. These are values that are outside of three standard deviations from the previous 12 months and might merit further investigation. An approach like this one can help flag outliers but might need to be done on a shorter timescale than is shown in this illustration.

Figure 3.3. Statistical Process Control Analysis of Depot Requisitions

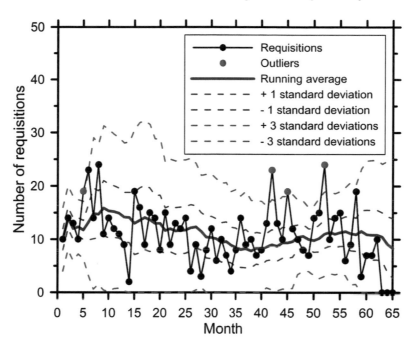

Note that all of these techniques draw attention to potentially anomalous data in cases when humans might not see the anomalous data or the anomalies are obscured in a very large, noisy data set. They flag data of interest to investigate further to determine whether the anomalies are problematic. The methods, therefore, *detect* anomalies but do not *evaluate* them.

Adjusting to Wartime

Wartime is not only the most important time to have effective measures to counter data corruptions by cyberattack, but it also challenges workers by differing from the peacetime environment to which they are accustomed. Regarding detection, the key difference between peacetime and wartime is that what appears anomalous in peacetime and wartime will sometimes differ. But the degree to which these times differ varies among the functional areas. While data for maintenance, such as technical data and maintenance procedures, will largely be the same at wartime, data for supply and transportation will differ considerably from peacetime. This change in baseline of normality means that detecting a true error during wartime becomes more difficult in these functional areas, just when errors can potentially be the most consequential.

Functional areas, such as supply and transportation, need to adjust to the wartime environment to effectively detect corrupt data. But what adjustments can be made? Relative to peacetime, wartime has both a higher threat of a cyberattack and greater consequences if action is taken based on false data. This change in risk suggests one way to address this challenge: For functional areas that are likely to experience large deviations between peacetime and wartime operational patterns, implement tiered response measures for detecting corrupt data.

A tiered response would be similar to the force protection condition (FPCON) levels used in accepting and mitigating risk of installation security. Based on threat, the Department of Defense defines five progressively increasing FPCON levels and associated specific protective measures: Normal, Alpha, Bravo, Charlie, and the highest level, Delta.[24] These levels are set and changed dynamically. For each FPCON level, specific protective measures are designed to meet the threat, yet not place an unnecessarily high burden on base operations or security forces personnel.

Detection of corrupt data could follow its own progressively tiered posture based on the separately assessed information operations condition (INFOCON).[25] Given the interconnected nature of cyberspace, the INFOCON is set based on global, not local, conditions. The point we want to emphasize is that both the FPCON and INFOCON levels prescribe specific measures that must be taken, given the specified threat level. The FPCON measures are levied to the security forces and the INFOCON measures to the cyber defense forces and network administrators.

We recommend that the logistics community define in policy what measures it should take based on INFOCON level. These measures would be aimed at operational continuity of logistics functions more so than at information assurance. Exactly what these measures might be and which would be applied to each functional area would require a separate study. Some examples of actions that could be taken include the following:

- Before taking any critical actions that would have a long recovery time if in error, verify data by separately contacting the appropriate authorities to confirm the prescribed action (e.g., calling the program office to verify the demilitarization of a part).
- Spot-check data and actively survey data integrity by other proactive means (to replace passive detection that would be common during normal operations).
- Replicate data more frequently in ways that are difficult to attack, such as printing out certain data (e.g., in-stock inventory and warehouse locations, technical orders) and copying them to isolated media (e.g., compact discs).
- Review and exercise manual and non–information technology procedures more frequently.

We recognize that moving to a tiered response will slow operations somewhat during higher cyber threat conditions, but it would also proactively hedge against longer delays or shutdowns

---

[24] Department of Defense Instruction 2000.16, 2006.

[25] INFOCON is soon to be replaced by a cyber conditions (CYBERCON) system. INFOCON levels are global and set by the Commander, U.S. Strategic Command. But commanders at all levels can set more-strict INFOCON levels locally for systems under their command (Chairman of the Joint Chiefs of Staff Instruction 6510.01F, February 9 2011, directive current as of June 9, 2015; Strategic Command Directive 527-1, January 27, 2006).

from the realization of a successful attack. Some slowdown is necessary for accurate detection. The Air Force increasingly uses and relies on data and information technology systems in order to reduce the workload on personnel. Not surprisingly, research has shown that the higher the workload and stress that workers have, the more they rely on data and the less likely they are to suspect or detect an error.[26] When dealing with a high workload, they are more likely to miscalibrate their trust and fall into the "excess trust" field of Figure 1.1. The same trade-off occurs when a base goes to a higher FPCON status—some efficiencies are sacrificed for an increase in security.

Having a tiered response would alleviate the need to expend resources during times of low threat and relatively low consequences of acting on false data. Changing procedures in the logistics community according to INFOCON would also serve as a deterrent to attack. Adversaries would know that the force is less susceptible to cyber attacks by virtue of adopting more robust and resilient measures.

Training for Detection

The degree to which people trust data and automation depends on their previous experiences with the accuracy of data and the usefulness of automation. The more reliable they perceive the data and automation to be, the more trust they tend to have, and the less likely they are to suspect (and, by inference, detect) problems.[27] Very trusting workers are at risk of acting uncritically on data and falling into the "excess trust" region in Figure 1.1. Workers with little trust are at risk of falling into the "excess suspicion" region in Figure 1.1 and taking actions against the direction of good data (such as replacing sound direction from data with their own judgment or ignoring a real alarm because they have heard many false alarms).

Referring again to Figure 1.1, to be most effective at detecting corrupt data, workers must stay on the "well-calibrated" line, expressing an appropriate level of trust given the real environment. Because the potential accuracy of data will vary depending on the threat environment—estimated by INFOCON—workers need to adjust their perceptions (trust in data) accordingly to remain well calibrated. But most of the perceptions of logistics workers arise from their training and day-to-day experiences, not wartime or attack experiences.

In general, these training and day-to-day experiences are unlikely to represent accurately what could transpire in a wartime situation involving an attack. It is unlikely that the U.S. Air Force has yet experienced the most capable attack through cyberspace, so workers have not experienced the full range of possible attacks. Training and exercises have been conducted with limited local effects, allowing units to reach out to other units for support. In a more far-reaching

---

[26] Endsley, 1995; Parasuraman and Riley, 1997.

[27] Lee and Moray, 1992; Parasuraman and Riley, 1997; Madhaven, Wiegmann, and Lacson, 2006; Merritt and Ilgen, 2008; Hoff and Bashir, 2015; de Vries, van den Berg, and Midden, 2015.

attack, all units might be at risk and might have no other unit available to reach out to for help. This lack of realistic experience limits the ability of logistics workers to conceive of what a targeted, coordinated cyber attack by a capable nation-state would look like. Training needs to fill this gap.

We did not conduct a systematic survey, but our anecdotal observations based on numerous meetings with logistics workers in the field spanning the major commands and at all ranks, from airmen first class to field-grade officers, confirm this lack of awareness. Every group with whom we met included individuals who were incredulous that the systems that they used could be attacked or that any actor could alter the data they used. Many workers were very confident in the abilities of U.S. Air Force computer security experts to keep their data free from adversary corruption. The real risks are not accurately reflected in these perceptions.[28]

On the other hand, workers, especially airmen, are subjected to a slew of safety-related briefings. Some relate to their personal time (e.g., motorcycle riding) and some to the workplace, the latter ranging from generic training regarding how to safely lift an object to specific training on avoiding injury when doing a particular maintenance action. Much of the workplace safety training is done by a supervisor in the place of work and includes clearly expressed consequences (which are sometimes graphic) of what can happen if safety protocols are not followed. The safety training received goes beyond the generic and is specifically documented regularly by the worker and supervisor using Air Force Form 55.

No parallel training in cybersecurity in the workplace exists. In fact, for most workers we encountered, they had no cybersecurity training beyond a briefing on operating a desktop computer. They are left with the impression that cybersecurity applies to the office environment but that the workplace (e.g., aircraft maintenance systems) is secure and they need not exercise any particular cyber hygiene there.

Without awareness of the real cybersecurity risks, workers are less likely to detect and report suspected corrupted data. Therefore, the principal recommendation for training is to educate all workers (airmen, civilians, and contractors) that corruption of data is possible and that they should be on the lookout for such aberrations.

A full syllabus of what this educational and training regime might look like would require a separate study. In its broad outlines, we envision education and training patterned after what is done to instruct workers regarding safety. The key objectives of education and training are as follows:

- Indicate that *all* data systems are susceptible to data corruption and cyber attack, and, therefore, vigilance is required for *all* systems and processes.
- Emphasize that everyone in the logistics enterprise has equities in cybersecurity, not just those with specific cybersecurity specialties.

---

[28] See, for example, Batey, 2016.

- Education and training should persist throughout a career, starting with technical schools and continuously thereafter during on-the-job training and exercises, and should be tracked with a separate record, similar to safety training and Air Force Form 55.
- Education and training should be performed, as much as practical, in the work environment with real examples and should emphasize possible consequences of failure to detect and report suspected anomalous data.
- Education and training should include all workers, including airmen, civilians, and contractors.
- Describe specific actions to be taken in a tiered posture by INFOCON, in the spirit of those outlined in the previous section, but in more detail.

Just like in safety training, the less abstract the message and the more concrete the needed actions by the worker and the potential consequences of not following those actions, the more effective the outcomes. Doing so in the cybersecurity domain will be challenged by the need to withhold information that workers are not cleared to know. Clearly, the more detail that can be given, the better insight the workers will have. But even the most general instruction that all data systems are susceptible to cyber attack has not been effectively communicated and would be beneficial. Specific examples of what could happen and the resulting consequences would help reinforce the message, but these can be notional.

A final observation about education and training is that some research suggests that younger cohorts of workers interact with data and automation differently than older cohorts. These differences appear to affect the degree of trust in data exhibited by each age cohort. That age is a factor in how people trust data is widely confirmed by research, but how different age cohorts behave appears to depend on context.[29] It might prove beneficial to tailor education and training for cybersecurity generally (and detecting data corruption specifically) according to age cohort. Unfortunately, the research on this age dependency is thin and does not yet lead to specific generalizations about how to instruct different age cohorts.

Even when logistics workers do detect some kind of error in data, there is currently little to no feedback to the workers on what happened and why. Was the error random? If so, how did it occur? Was it an attack? In the absence of such feedback, workers lack key information needed for them to calibrate their trust. It has been long recognized in the field of safety that such feedback is critical for a satisfactory safety culture.[30] When a near miss or accident occurs, formal reporting aggregates and assesses the information and disseminates lessons learned to workers via reports, altered training, and other mechanisms. Feedback on performance is also needed for effective monitoring.[31] Effective feedback requires a workforce motivated to report, formal mechanisms for them to report, rapid evaluation capabilities to extract lessons, and ways to disseminate those lessons back to the workforce. We take up the topic of motivating the

---

[29] Hoff and Bashir, 2015, and references therein.

[30] Reason, 1998; Woods, 2006; Dekker, 2006. See Air Force Instruction 90-802, 2013.

[31] Parasuraman and Riley, 1997.

workforce in the next section. The other topics are, to varying degrees of detail, addressed in the next chapter.

Motivating Detection

The literature on how to motivate workers is vast, and it is clear that there is no silver bullet that works in all settings. We focus here on some key insights from the psychology and social science literatures that are particularly relevant to the problem of motivating workers to report suspected errors in data. In managing the motivation of a workforce, leaders need to address two questions. The first key question is: What do leaders want workers to be motivated to do? The second key question is: What are the most effective mechanisms to motivate workers, given those desired outcomes?[32] We address the first question from a broader view of shifting the culture in the workforce. This discussion is patterned after how safety culture has developed but has not yet developed in the same way for cybersecurity.[33] Given these broad institutional directions, we then discuss mechanisms to motivate individuals to achieve these objectives.

Creating and Sustaining a Reporting Culture

Like safety, there is a general need to create and sustain a *culture* for cybersecurity that encourages workers of all types to report suspected anomalous data. Schein argues that:

> *Culture* can . . . be defined as (a) a pattern of basic assumptions, (b) invented, discovered, or developed by a given group, (c) as it learns to cope with its problems of external adaptation and internal integration, (d) that has worked well enough to be considered valid and, therefore (e) is to be taught to new members as the (f) correct way to perceive, think, and feel in relation to those problems. The strength and degree of internal consistency of a culture are, therefore, a function of the stability of the group, the length of time the group has existed, the intensity of the group's experiences of learning, the mechanisms by which the learning has taken place (i.e., positive reinforcement or avoidance conditioning), and the strength and clarity of the assumptions held by the founders and leaders of the group.[34]

For the problem of creating a culture for detecting and reporting suspected errors in data, this argument naturally leads to the need for (1) strong leadership, (2) protection for workers who report, (3) feedback on performance, and (4) clear guidance on what actions to take. We discuss recommendations for each of these in more detail.

Leadership. Leaders in the logistics community need to make data integrity a priority and need to communicate that priority to workers. Leadership is needed at all levels. At the lower levels, supervisors need to show that reporting is important to them and their superiors. At the

---

[32] Higgins, 2012, especially Chapter 12.

[33] Reason, 1998; Woods, 2006.

[34] Schein, 1990, p. 111. Italics are in the original.

highest levels, leaders need to express the importance of the integrity of data and, therefore, the need to detect and report any suspected issues. It is a countermotivating factor that many errors in data are human errors. Humans and organizations have a tendency not to want to report their mistakes. Both tend to try to resolve any issues with mistakes before elevating the issue to an outsider. If anomalous data might be due to human error, this tendency leads to a countermotivation to reporting. The very act of reporting and evaluating suspected issues in data will inevitably slow down the work pace somewhat. These are potential negative repercussions of reporting. Workers need to know that leaders are behind them.

Leaders must also emphasize that all members of the organization have equities in cybersecurity in general and detection of data corruption in particular. It has been long accepted that all members of an organization play a role in safety. Safety, or the lack thereof, emerges out of the collective actions of all members of an organization. It is not the unique responsibility of a special group of safety experts in a lead organization. Cybersecurity shares this quality. The cybersecurity of an organization emerges from the collective behavior of all of its members. Only effective leadership can instill this shift in attitudes.

Protection. Workers need to be protected from any retaliation when they report.[35] This protection might require leaders to issue policies that explicitly protect reporters, keep the identity of reporters confidential, or both. Confidentiality is central to many successful reporting systems in safety. It is a keystone of the Aviation Safety Reporting System (ASRS),[36] which is one of the most successful safety programs and is widely credited for the extraordinary safety record of commercial aviation. The inception of the ASRS coincided with an enormous drop in the number of commercial aviation accidents.[37] Confidential reporting of issues within the commercial aviation sector is protected by law (unless the reporter discloses unlawful behavior). As a corollary to protection, it might be beneficial in some cases to administer punishment or admonishment for workers who knowingly fail to report an error of consequence.

Feedback. Workers need feedback on their performance in detecting and reporting. Feedback is important both to help workers learn to better calibrate their trust in data and to motivate them to improve how well they discharge their jobs. In our framework, we consider feedback as part of the evaluation process and discuss it further in the next chapter.

Guidance. Workers need to have clear guidance on what actions they are expected to take. What is appropriate to report, and when? To whom? Workers need a clear process and formal channels for reporting. Reporting processes for suspected erroneous data need to be easy and should mimic other reporting processes as much as possible.[38] The limits of what they are to do must also be made clear to workers. For example, it is the role of the evaluation group(s) to

---

[35] Weick and Sutcliffe, 2001, Chapter 5.

[36] Reynard et al., 1986; see ASRS, undated.

[37] Dijkstra, 2006.

[38] Thaler and Sunstein, 2008.

diagnose the problem and to instruct on mitigations, not the worker. We take up some of these processes in the next chapter.

Given these desired cultural shifts, what guidance does the literature provide for motivating individual workers along these lines?

Mechanisms for Motivating Individuals

Recent research in psychology suggests that the effectiveness of various mechanisms for motivating people depends on the degree of match between the orientation of the motivation mechanism and the manner in which a worker performs a task.[39] No all-purpose tool, such as introducing incentives (carrots or sticks), effectively motivates workers. Effective motivation emerges from a mutual fit among the state of the worker, the task to be accomplished, and the mechanisms employed to motivate that worker. To understand the academic thinking in this area of motivation, some background is necessary.

Psychologists define two states of workers in this setting: *regulatory focus* and *self-regulatory mode*. We first consider the two kinds of regulatory focus: *promotion* and *prevention*.[40] In a classroom, a student with a promotion focus would aspire to a grade of an A and would view an A as an accomplishment. A student with a prevention focus would feel a responsibility to get an A because the student felt that he or she ought to do so and would fear any other outcome. The promotion focus emphasizes advancement, accomplishment, and attaining new objectives. The prevention focus emphasizes safety, security, avoiding loss, retaining what is already possessed, and preserving the status quo.[41]

Although individuals might have a general proclivity toward having a promotion or prevention focus, a single worker can present a promotion focus in one situation and a prevention focus in another. If the worker perceives that the status quo is unacceptable in a certain task, the worker might adopt a promotion focus. If the task is to clean up a workspace, for example, the worker would perceive the status quo as unacceptable and strive to create a better workspace, placing the worker in a promotion focus. On the other hand, if the worker perceives the status quo to be good and there is danger in disrupting it, the worker might adopt a prevention focus. This might be the case if a maintenance worker were reluctant to adjust a machine that appeared to be working fine.

Which regulatory focus a worker adopts can be manipulated (advertisers take advantage of this fact). Experiments indicate that a person's regulatory focus can be altered by merely changing the way in which a problem is framed. In one example, those in a study who were told that getting a certain fraction of a task correct would get them a fun second task (reward) tended

---

[39] Most of the arguments in this section derive from Higgins, 2012, and Cesario, Higgins, and Scholer, 2008. See also Tyler and Blader, 2005.

[40] Higgins, 1996.

[41] Crowe and Higgins, 1997.

to adopt a promotion focus. Those who were told that getting a certain fraction wrong would get them a boring second task (punishment) tended to adopt a prevention regulatory focus.[42]

The second state of workers is their self-regulatory mode. Self-regulatory modes are strategies selected to match the focus. An *assessment self-regulatory mode* is one in which an individual wants to assess all alternatives before committing to an action, even if it leads to delays in the action. This person is vigilant, analytical, and wants to "get it right." A *locomotion self-regulatory mode* is one in which an individual wants to take action, even if it means not considering all the possible alternatives at their disposal. This person is eager, enthusiastic, and wants to "get it done."[43] The locomotion strategy is a *regulatory fit* to the promotion focus; the assessment strategy is a regulatory fit to the prevention focus.[44]

A person's strength of engagement is greater in cases when there is a regulatory fit with the task objectives that they have.[45] This observation leads us to how to most effectively motivate workers. *Motivational mechanisms should fit the desired regulatory focus and self-regulatory mode and should be framed in a way that nudges workers into the desired regulatory focus, given the work goals.* Working through two examples helps to illustrate this point; the argument is summarized in Table 3.1.

Table 3.1. Summary of Motivational Framework

	Domain	Environment A	Environment B
Management objectives	Goals	Quality	Quantity
		"Get it right"	"Get it done"
		Reflection, analysis	Production
		Vigilance	Action
	Regulatory focus	Prevention	Promotion
	Self-regulatory mode	Assessment	Locomotion
How to motivate	Incentives	Non-loss rewards	Gaining rewards
		Punishments	
	Leadership messaging	Obligations	Pride
		Responsibility	Eagerness
	Environment	Serious	Fun
	Effective feedback	Negative	Positive

SOURCE: Summarized from Higgins, 2012.

---

[42] Roney, Higgins, and Shah, 1995.

[43] Higgins, 2012, pp. 130–133.

[44] Higgins, 2012, p. 87.

[45] Higgins, 2012, p. 87.

For the first example—Environment A in Table 3.1—consider a task that demands quality over quantity. This is a case where workers should display reflection, undertake analysis, and be vigilant. They need to "get it right." This situation fits a prevention regulatory focus, and the appropriate self-regulatory mode is assessment. To motivate workers in this situation, motivational mechanisms should have a regulatory fit to these goals.[46] If incentives are used, they are best cast in a non-loss form rather than as a reward to be gained. This means telling workers that they will have a reward of some kind *so long as they do not fail in the task* rather than that they will get a reward *if they succeed in a task*. Punishments are also a regulatory fit, but they could impede motivation if they are perceived as unfair. Beyond incentives, messages from leaders are a better fit if they are cast in terms of obligations and responsibilities rather than in terms that elicit pride and eagerness.[47] The overall environment is a better fit if it is serious rather than fun. And, finally, feedback is a better fit if it is negative (reporting on failures) rather than positive (reporting on successes).[48]

In the second example—Environment B in Table 3.1—quantity is valued over quality. In this case, workers should emphasize production and action; they need to "get it done." This situation fits a promotion focus, and the appropriate self-regulatory mode is locomotion. To motivate workers in this situation, methods fit best that are the opposite of the first case. If incentives are used, they are best cast in the form of a reward to be gained rather than in a non-loss form. Punishments are a non-fit and should be avoided. Beyond incentives, messaging from leaders of the goals are a better fit if they are energetic and are cast in ways to encourage pride and eagerness, not in terms of obligations and responsibilities. The overall environment is a better fit if it is fun rather than serious. And, finally, feedback is a better fit if it is positive (reporting on successes) rather than negative (reporting on failures).[49]

Motivating Workers to Detect Data Anomalies

What does this mean for motivating the detection of anomalous data in logistics? Given the goals of this study, U.S. Air Force logistics workers (airmen, civilians, and contractors) should be ultimately motivated to provide robust and resilient logistics operations and proximately motivated to report any data anomalies that they detect that could harm logistics operations. These objectives are consistent with a workforce that is attentive, reflective, and vigilant. This workforce should prefer accuracy over quantity of production. It is a workforce that might have the motto of "get it right." Therefore, to motivate the workforce, incentives should be cast as non-loss rewards, messaging should be in the form of obligations and responsibilities, the tone of

---

[46] Higgins, 2012, pp. 239–242.

[47] Cesario, Higgins, and Scholer, 2008.

[48] Higgins, 2012.

[49] Higgins, 2012.

messages and the work environment set by supervisors should be serious, and the feedback should be in terms of reporting failures.

Following these principles is not a magic formula for enhanced detection of data anomalies. These are trends suggested by research in what motivates people. The patterns summarized in Table 3.1 are well established in the literature and work in practice. An uncertainty in their application is the degree to which workers should be induced to fall into prevention or promotion roles. Leadership styles that mediate followers' regulatory focus have mixed results in safety.[50] Both a promotion and a prevention focus can enhance safety. Promotion encourages proactive thinking about avoiding accidents and improving habits but diminishes compliance rates with safety rules. Prevention does the opposite. The outcomes of application of these principles to data detection should be monitored and the methods adjusted as appropriate.

Note also that these objectives are inconsistent with a workforce that is focused on high production rates and constant activity directed toward production outcomes—in short, one with a motto of "get it done." Both of these workforce foci—"get it right" and "get it done"—are desired in U.S. Air Force logistics, but they create a tension in the workforce. All other things equal, a "getting it right" environment is expected to yield more, and more accurate, reporting of anomalous data. It is also consistent with a tiered response framework for cyber threat condition levels (INFOCON) in which efficiency is traded for security. On the other hand, a "getting it done" environment is expected to inhibit such reporting. Air Force leaders will want to tailor the motivational mechanisms used to the case of anomalous data detection or, at the very least, be careful not to interfere with other cases in which a promotion focus is preferred.

Conclusions and Discussion

Many of the recommendations proposed in this chapter recur in discussing how to address multiple challenges. To put more systematic order to these recommendations, in this section we gather the key recommendations together into a single list in approximate order of importance by general category.

Education, Training, and Leadership Messaging

We assess that the foremost recommendation is to educate and train all members of the enterprise. The key elements of the message are as follows:

- Educate workers at all levels that *all* data systems are susceptible to data corruption and cyber attack, and, therefore, vigilance is required for *all* systems and processes.
- Instill a culture in the workforce that everyone in the logistics enterprise has equities in cybersecurity, not just those with specific cybersecurity specialties.

---

[50] Kark, Katz-Navon, and Delegach, 2015; Aryee and Hsiung, 2016.

Each level of leadership (command and supervisory) should play a role in this messaging. Senior levels (generally flag- and field-grade officers) should set the reporting of cybersecurity concerns as a priority. They should make that priority concrete by extending protection from retaliation to reporters. And they should write policy to ensure that reporting is easy to do and that guidance is specific. The goal is to shift the culture so that workers are as vigilant in reporting suspected corrupted data and other cyber concerns as they are with foreign object damage control on a flightline and other safety concerns. Currently, there is little to no reference to data integrity or cybersecurity in logistics issuances.[51]

Lower-level leadership (generally company-grade officers and senior noncommissioned officers) should develop clear, specific guidance for workers on what kinds of data anomalies should be brought to their attention, based on their knowledge of operational needs.

Messaging should extend to all workers, to include airmen, civilians, and contractors. It should persist over time. For airmen, it should start with technical schools and continue thereafter during on-the-job training and exercises. And it should be performed, as much as practical, in the work environment with real examples and should emphasize possible consequences of failures to detect and report suspected anomalous data.

The tenor of the messaging at all levels is best done when delivered as obligations and responsibilities to detect and report. Rather than issuing rewards to be gained for accurate detection and prompt reporting, non-loss rewards should be issued for success and appropriate punishment for failures to report. These messages are best delivered in a serious rather than fun work atmosphere. And feedback to workers on their collective performance at detection and reporting will be most effective when it emphasizes failures and their consequences more so than successes.

The area of safety can be used as an exemplar of how to do this education and training. These challenges and the proposed recommendations share many attributes with safety. Safety is a concern in all environments. All workers have equities in safety, not just those in safety organizations. Leaders at all levels make it a priority. Education and training for safety is given to all workers persistently over their careers. Guidance is as specific as possible. Real examples emphasizing negative outcomes if safety procedures are not followed are used as much as possible. Reporting of safety concerns and violations is encouraged and expected. Often, confidentiality is extended. Non-loss rewards for success and punishment for failure have been successful. Cybersecurity can advance considerably by emulating these habits. To be clear, we are not recommending merging safety and cybersecurity, just using some of the same successful solutions to address similar challenges across the two areas.

---

[51] For example, the foundation policy for aircraft maintenance, Air Force Instruction 21-101, 2015, does not convey these messages.

Posture

We recommend that the logistics community define in policy what measures it should take in response to each INFOCON level. Of all the recommendations in this report, we assess that this action is most likely to yield positive, tangible results with the least expenditure of resources. These measures would be aimed at operational continuity of logistics functions more so than at information assurance. Procedures within logistics would be specified to parallel existing procedures specified for information technology systems. For each INFOCON level, specific proactive measures would be defined in each functional area for increasing robustness and resiliency. These measures would progress with increasing threat so as to match the expended resources (and any costs to efficiency) with the threat environment. Just as in FPCON levels, the measures to be taken would be specific and tailored, and they would be written into U.S. Air Force Instructions. Some examples of actions that could be taken include the following:

- Before taking any critical actions that would have a long recovery time if in error, verify data by separately contacting the appropriate authorities to confirm the prescribed action (e.g., calling the program office to verify the demilitarization of a part).
- Spot-check data and actively survey data integrity by other proactive means (to replace passive detection that would be common during normal operations).
- Replicate data more frequently in ways that are difficult to attack, such as printing out certain data (e.g., in-stock inventory and warehouse locations, technical orders) and copying them to isolated media (e.g., compact discs).
- Review and exercise manual and non–information technology procedures more frequently.

Tools

To the extent practicable, the U.S. Air Force should consider employing standard techniques used in fraud detection to detect data corruption in data sets that humans do not see during normal operations. These tools should be preferentially used on functions and data identified as priorities by the framework described in Chapter Five. Benford's Law is one relatively easy check on data, but other standard methods could also be used. The extent of use of these techniques should be set according to the cyber threat condition to economize on workload.

4. Evaluation

Once a suspected data error has been detected, it needs to be reported to someone who can evaluate it. Is it actually an error? If so, is it a normal mistake, or was it a malicious attack? This step has two key phases. First, workers need channels and standard processes for reporting. These were mentioned in the previous chapter and will be further developed in this chapter. Second, the enterprise needs an organization to receive the reports, interpret them, and disseminate any findings. The error evaluation has roles and implications for both logistics and cyber operations.

In this chapter, we begin by outlining the needs of both reporting and evaluation organizations. We then compare those needs with existing processes and organizations in the U.S. Air Force and examine how these can be leveraged to address the problem of robust and resilient logistics operations in the face of cyber attack. We close by identifying gaps in existing reporting and evaluation and by proposing some modest additions to existing processes and organizations that would enhance robust and resilient logistics operations in the face of compromised data.

The Needs for Reporting and Evaluation

Reporting

Chapters Two and Three described the key desired characteristics of reporting processes. Reporting needs to be (1) comprehensive, (2) timely, (3) easy to do, (4), available to all workers, and (5) protected from recriminations from reporting.

The process, or processes, must comprehensively span all potential data corruption. Reporting mechanisms for specific types of data within a given functional area are very useful. But there must be some reporting mechanism available to any worker who notices data corruption in any database.

The process must be timely. Timeliness requires prompt action on behalf of the reporter and quick transmission through the reporting channel. If the process must pass through multiple layers of bureaucracy—for example, requiring it to be approved through several levels in the chain of command—time will be lost at each step. And each step provides an opportunity for someone to deliberately or inadvertently stall or intercept the report. The report needs to pass directly to an evaluator, notifying others only for situational awareness.

The process must be easy. Workers will tend to delay or avoid doing a burdensome task that could go unnoticed if they failed to do it. Onerous reporting, either for the worker or for any person in the transmission channel to the evaluator(s), will delay response. Reporting can be onerous in several ways. It can require a lot of paperwork on behalf of the worker, or it can place a high knowledge burden on the worker—for example, by requiring the worker to choose among many reporting systems or to know in advance what organization is best positioned to assess the report. The effects of a burden on reporting will be amplified during times of high workload, such as during high operations tempo, just when an attack is more likely and of more concern. Reporting is best done with minimum paperwork, simple instructions for submission, and a process that is familiar to the participants (for example, mimicking similar processes).

The process must be available to and effective for all workers. Because data errors can occur anywhere, whoever is best positioned to detect and report them needs to have a reporting system at hand. A single reporting system must be able to accommodate airmen, civilians, and contractors.

The process should protect the reporter. Potential reporters might be the only ones with the knowledge to report, so failing to report would not, in these cases, place them at any jeopardy. On the other hand, if they fear retribution for reporting, many will probably choose not to do so. This asymmetry discourages reporting. Protection from retribution for reporting will need to come from leadership priorities and messaging. As argued in the last chapter, those changes will emerge from alterations to workplace culture. But because the reporter might be an airman, a civilian, or a contractor, and these groups operate under very different oversight, confidentiality might be needed to provide a sufficient guarantee of freedom to report. Some safety reporting systems use confidentiality of reporting as a key component of garnering critical reporting.[52]

Evaluation

Evaluation entails numerous activities, the most salient being (1) prioritizing the order for assessing incidents and establishing how quickly each needs to be done; (2) assigning each incident to an assessor or assessors; (3) distributing the incidents to assessors and any other organizations for situational awareness; (4) assessing incidents; (5) disseminating findings from the assessments; and (6) compiling, archiving, and disseminating lessons learned. Each of these activities will need to be done in the logistics (combat support) realm and in the cyber operations realm. We will focus most of our discussion on where and how these activities can be done within logistics (combat support).

Prioritizing. In any reporting system, assessors will receive numerous incidents, and they will need to prioritize assessing some over others. Each assessment itself will need to be done within some time frame, preferably before operators make critical decisions based on the data. Therefore, the first organization to receive reports should be equipped to determine how quickly an incident should be assessed and the priority of assessing incidents. Both the urgency and priority can be judged using the framework presented in Chapter Five. In a large volume of reports, some might be easily judged in a preliminary screening to be unworthy of any detailed

---

[52] The ASRS is an example.

assessment because of complete or partial duplication in the assessment pipeline or obvious lack of significance.

It is important that the organization prioritizing incidents be a central authority. Only when a single organization sees every report can it adequately judge relative priority. Another reason for a central clearinghouse is to have the perspective to discover potential correlations among the reported issues. Consider a coordinated attack on a single weapon system that unfolds such that all of the following areas experience a small but significant attack: flightline maintenance, supply chain support, local base operations, and supporting transportation. Without a central authority, each of these small attacks might be assessed as being an inconsequential nuisance. But with a central view, they would be recognized as a coordinated assault on a single weapon system. Authorities determining a plan for response and mitigation need that broad perspective.

Assigning. After prioritizing and determining how quickly an assessment is needed, each report needs to be assigned to a group that has the requisite expertise to assess the incident. The organization making this assignment must have enough knowledge to know where to route the incident and must be empowered to task the receiving organization.

Distributing. The incident, its priority, and the timeline for assessment then must be distributed to the assessment organization(s) and any organizations that should be aware of the incident for situational awareness.

Assessing. Assessment requires specific technical knowledge in the functional area. This knowledge is dispersed in many organizations throughout the enterprise, including program offices, functional operators, the Defense Logistics Agency, contractors, and many others. Hence, assessments will need to be dispersed. For each incident, the assessment activity would be assigned to an organization where the relevant locus of knowledge resides. That assessing organization would then determine whether the data were indeed in error. This part of the assessment requires both a logistics and a cyber viewpoint, and so the evaluation bifurcates at this stage.

On the logistics side, assessors will first need to determine whether the incident reported is actually an error. For example, a reporter might see a change in technical data or routing of a supply part and suspect that it was corrupted. However, the assessor, after some investigation, might discover that the data were in fact accurate. Or the assessor might determine that a data entry error or some other routine mistake had occurred.[53] If the data were assessed to be corrupt and no clear explanation was evident, then the incident would be shared with an assessor on the cyber operations side to determine whether the corruption was caused by a cyber attack.

---

[53] Many databases contain log files that record which user made a change to the data at what time. It is important for the assessor to understand that a change made by an authorized account may still be corrupted data from a cyber attack and not a routine error. A key part of the assessment is to determine the root cause of the data corruption, which will usually require talking to the workers involved, not simply accepting the log file as correct. A sophisticated cyber actor, once in a database, can alter a log file.

On the cyber operations side, after assessors determine whether the incident appears to be a cyber attack, they will need to assess whether it is correlated with any other incidents and what remedies to direct. When this assessor ascertains that an incident is an error and that the error was due to a deliberate cyber attack, operational security concerns might restrict the dissemination of that assessment to a limited number of cleared individuals. This security need will then restrict the closure of the feedback loop to the reporter. The assessor might nevertheless pass along instructions to isolate computing equipment, lock down systems for forensic investigations, or take other actions.

Disseminating. The assessments then should be reported back to the central authority, the same authority that assigned the assessment.

Compiling. Some organization should compile, archive, and disseminate any lessons learned from these incidents. If some common mistake is causing multiple data errors, these lessons can be used to modify policies, training, and processes to reduce future data errors.

Existing Reporting Channels and Their Limitations

During normal operations, data errors of various kinds occur in every logistics function. Consequently, every logistics function has evolved one or more ways to perform some of these reporting and evaluation activities. We briefly review some of these activities to explore what can be exploited for logistics robustness and resiliency and what additional capabilities are needed to augment existing reporting channels and evaluation processes.

One of the most common—and nearly ubiquitous—reporting mechanisms is some form of help desk or call center. These are organizations that can be contacted, usually by email and/or phone, to resolve data discrepancies and other complaints. These services often operate during normal duty hours and sometimes 24 hours a day, 7 days a week, to address emergency and urgent matters. Most large, authoritative information technology systems have some kind of help desk or call center. Functional areas also have similar processes, especially for addressing data discrepancies. Supply, for example, has a formal process for reporting supply discrepancy reports via Standard Form 364.[54] The fuels functional area has a help desk to report automated information system trouble tickets.[55] There are numerous other examples.

In addition to help desks and call centers, functional areas such as aircraft maintenance have developed multiple processes for reporting and verifying suspected errors in technical data. Maintainers who question technical data can report discrepancies to program offices via a Form 22 process.[56] These are routed from the worker, through the wing, to the major command, and

---

[54] Air Force Instruction 23-101, 2016, Section 5.9.9.5; Air Force Instruction 21-201, 2015, Section 7.9.3.

[55] Air Force Instruction 23-201, 2014, Section 6.10.

[56] Technical Manual TO 00-5-1, 2014, Section 9.

then to the originating organization responsible for the technical order (e.g., a program office).[57] When technical order data do not adequately address a problem, maintainers have a separate process to send technical assistance requests to those with engineering expertise, colloquially referred to as "107 requests," named after the associated technical order.[58] Both of these reporting mechanisms are formal processes familiar to all maintenance workers that provide means for reporting specific concerns regarding data integrity.

It seems a fair generalization that the more critical the potential data error, the more formal and more well developed the reporting and evaluation mechanisms have become in individual logistics functional areas. In supply and transportation, where individual errors are relatively common and generally not immediately critical to operations, reporting mechanisms and evaluation procedures are mainly limited to call centers and direct communication with the involved parties in the transaction. In maintenance and explosive ordnance disposal, where technical data must be accurate and precise to avoid grave consequences, multiple mechanisms have been developed. These are very formally defined procedures with assigned accountability.

Areas in which data errors could affect flight safety or aircraft availability—such as aircraft maintenance technical data and aircraft parts—often include in the reporting mechanisms some form of prioritization and expected timeline for resolution. Spare parts orders, for example, are given mission impact codes and bear a system reporting designator so that parts with a mission-capable (MICAP) condition can be indicated and filled as a priority.[59] Form 22 reports are assigned emergency, urgent, or routine prioritization status. Similarly, 107 requests are assigned a status of routine, urgent, immediate, or emergency. These categories guide the order in which to assess reports and, often, time frames for responding.

We have seen that there are a number of existing mechanisms that can help with reporting suspected data corruption. Most of these are not onerous, and their use is part of standard training. For the purposes of robust and resilient logistics operations in the face of a cyber attack, however, they do not meet all of the criteria listed above in this chapter. We have identified a number of gaps. Existing evaluation mechanisms

- do not provide comprehensive means for workers to report all possible data corruption incidents. They are restricted to specific data types, data systems, and functional areas.
- place the burden on the worker to identify the proper reporting mechanism for any given incident
- are not available to all workers, as many are specific to airmen in particular jobs and are not available to all civilians and contractors who come into contact with data and are potential reporters
- do not provide an enterprise view of issues that could identify a coordinated attack that cuts across functional areas or organizations

---

[57] Technical Manual TO 00-5-1, 2014, Section 9.6.

[58] Technical Manual TO 00-25-107, 2011.

[59] Air Force Instruction 23-101, 2016.

- do not fully assess the causality of an incident, archive these lessons, and disseminate that information systematically to revise policies, education, and training
- are not, in general, timely enough for responding to all types of potential attacks
- do not comprehensively assign a priority to each incident and how timely each response should be
- place, in a few cases, an unnecessary burden on workers to report. For example, a worker who questions technical data originating in the Air Force must file a Form 22 request. But if those technical data originate with a contractor, the process is different and requires writing a letter through the major command to the appropriate program office. The nature of the response back from the contractor depends on the specific contract language under which they operate.[60] This process could be simplified by providing one reporting mechanism for the worker and placing the burden for routing the assessment to the program office (or other recipient).

A Proposed Response Structure for Reporting and Evaluating

The U.S. Air Force can leverage the reporting mechanisms and assessments that are already in use, some of which were described in the previous section, to support the broader effort to be more robust and resilient. Supplementing these with a minimal number of additional processes and organizations would largely close the evaluation gaps that we have identified. The proposed overall structure is shown in Figure 4.1 and is described in the remainder of this section. We divide roles and responsibilities between the logistics (combat support) and the cyber operations realms, specifying more detail on the logistics side.

Logistics Operations

We begin with the left side of Figure 4.1, which treats the evaluation of logistics operations. In Figure 4.1, existing structural elements (reporting channels and organizations) are depicted in black and blue and newly proposed structural elements in red. Reading the figure from the bottom up, airmen, civilians, and contractors (denoted by the box labeled "Reporters") report suspected incidents of corrupted data. The many existing channels (e.g., help desks, Form 22 reports, 107 requests) are shown by the black line. These reports go to various organizations for some form of assessment (e.g., program offices, call centers, the Defense Logistics Agency) and often are routed through or reported to other organizations for situational awareness (e.g., major command staffs), as shown in the box to the right of the "Situational awareness" bracket. After assessment, some feedback generally flows back to the reporters. In a few cases, the feedback might result in the change of policy, education, or training, but remedies of that kind are not currently systematically performed.

No authority has visibility of all of these currently performed reports in order to assess whether a coordinated attack might be taking place. Nor is there a central authority accountable

---

[60] Technical Manual TO 00-5-1, 2014, Section 9.

for determining the cause of any significant corruption of critical data and disseminating causal information for remediation.

Figure 4.1. Proposed Response Structure for Reporting and Evaluating

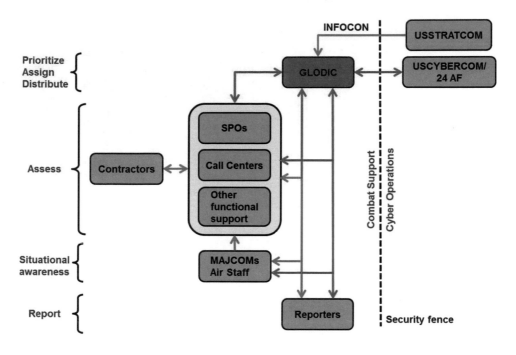

NOTES: Lines in blue indicate existing channels. Lines and boxes in red indicate proposed new channels and entities. GLODIC = Global Data Integrity Cell; MAJCOM = major command; SPO = system program office; USCYBERCOM = U.S. Cyber Command; USSTRATCOM = U.S. Strategic Command.

Hence, we propose a new central body that we call a *Global Data Integrity Cell* (GLODIC) that would receive copies of all such reports. This cell would maintain enterprise-wide situational awareness of all existing reports. Because these reports are being acted upon by the assessors, defined in the policies governing the current reporting mechanisms, most of the role of the GLODIC for these reports would be to look for trends and potential coordinated attacks, extract lessons learned over time, and serve as a clearinghouse of these lessons learned.

Because the existing reports do not span all possible incidents and are not available to all workers, especially contractors, a supplemental reporting mechanism is needed to close this gap. This mechanism would be dedicated to reporting suspected corrupted data and other potential cyber attacks; it is depicted by the red line connecting reporters with the GLODIC. The GLODIC would first do a triage. Some reports might be spurious or duplicative of existing reports. The GLODIC would prioritize the reports that pass triage, determine how rapidly they need to be assessed, assign them to a lead assessor, and distribute the report to any other organization with a need-to-know for situational awareness. The lead assessor could also draw from the full range of organizations with requisite knowledge, to include other U.S. Air Force organizations, contractors, and other agencies (e.g., Defense Logistics Agency). The main goal of assessment is to determine whether the incident is, in fact, an error, and, if so, whether the error is a normal mistake or a potential attack. That is the main information that the GLODIC must track.

To close as many reporting gaps as possible, this channel should be easy to use, perhaps being modeled after the maintenance community's 107 process, and should protect the confidentiality of the reporter. The latter would encourage participation of contractors and members of the U.S. Air Force who might fear retribution from those higher in their chain of command. The assessor could either send assessments directly back to the reporter, copying the GLODIC, or have the GLODIC send the assessment to the reporter.

The GLODIC needs to have enough expertise to do the triage, prioritization, and assignment of lead assessor. It must also have enough institutional knowledge to be aware of what organizations need to be copied for situational awareness. And it needs to be positioned such that it can reasonably be empowered to task the assessors and receive timely, effective responses.

No single organization meets all of these criteria perfectly, but one possible location for the GLODIC is the Air Force Materiel Command Directorate of Engineering and Technical Management (AFMC/EN). We expect that most of the assessments would be performed in the program offices under the Air Force Life Cycle Management Center, at the Air Force Support Center, or the Defense Logistics Agency. AFMC/EN is well positioned to make the first two assignments. Further, the weapon system cybersecurity expertise that would aid in prioritization and lessons learned is concentrated in the Air Force Materiel Command.

Whatever organization takes on the role of the GLODIC, it will need to increase its manning levels. It was beyond the scope of this study to assess the skills and quantify the manpower levels needed for such an organization, but the overall breadth of the activities assigned to the GLODIC in combination with the needed timeliness of response would not be achievable by adding new tasks to an existing organization.

Organizations with a cyber mission do not appear to be good candidates because the role of the GLODIC is mission oriented, not information security oriented. A priority is the continuation of the mission, not just integrity of the data, and this mission view must come from the logistics operations side. An information system security manager has responsibility for improving cybersecurity and restoring operations after a cyber failure.[61] Yet information security is a key piece of evaluation. Information security is not the focus of this report, but some discussion is needed for completeness.

Cyber Operations

Our discussion here of the evaluation process on the cyber operations side is necessarily incomplete. We focus on those elements needed to directly support the functions of the proposed

---

[61] Air Force Instruction 33-210, 2008, Incorporating Change 1, October 2, 2014; Technical Manual TO 33-1-38, 2015.

GLODIC. The first proposed supporting activity is the issuance of the INFOCON status. The INFOCON level issued by the Commander, U.S. Strategic Command, is global, but any commander can issue a stricter INFOCON level for any systems under his or her command. As described in Chapter Three, the INFOCON level would trigger a tiered response structure for reporting, assessment timelines, and mitigations *within the logistics community*. The INFOCON would flow to the GLODIC, which would disseminate it to all workers. Each functional area would define in policy how it will adjust operations for each INFOCON level. We emphasize that the only action taken by the cyber community is setting the INFOCON level; all other actions are done by the logistics community according to policies set by the logistics community on how to respond to various INFOCON levels.

The second proposed supporting activity is to assess whether a cyber attack has occurred. The assessment should be performed according to where the requisite expertise lies and does not have to be conducted by a central organization. For weapon systems, the locus of expertise might lie in the Air Force Materiel Command. For networked information technology systems, the locus of expertise might lie in 24th Air Force. But, as with the GLODIC, some organization needs to form an enterprise view, archive incidents, and disseminate instructions. Instructions might include whether to isolate systems from networks, change procedures, and lock systems for forensics. We call this organization the *Global Information Assurance Cell* (GLOIAC). It does not need to receive all reports, just those determined to be potential cyber attacks.

Summary

Our proposed structure for evaluation meets the five criteria of (1) comprehensiveness, (2) timeliness, (3) ease of use, (4) protection of workers from recriminations via confidentiality, and (5) availability to all workers. It fills the gaps in the current reporting and assessment mechanisms and does so economically by supplementing existing processes and organizations with minimal additions—a GLODIC—for prioritizing, assigning, and extracting lessons learned from incidents and a confidential channel modeled after 107 requests for reporting suspected cyber matters.

5. Prioritizing the Effort

Given the vast amount of data that could be in error and the large number of logistics functions that these data support, the logistics community needs some way of prioritizing efforts for detecting, evaluating, and mitigating the potential negative effects of anomalous data. In this chapter, we turn to the question of where attention to anomalous data is most vital and where risk might be accepted.

What to Include in Prioritization

For some data errors, avenues to mitigate errors might already exist. Workers might circumvent the data by direct communication with another unit or check the data against a reliable source, perhaps a paper record. One factor that might be included in ranking for risk prioritization is the availability of such preexisting mitigations. For example, if potential mitigations were considered in prioritization, a potential data error that could lead to a significant negative operational consequence might be considered a low risk when mitigations are available. On the other hand, possible mitigations could be excluded from the prioritization and considered after the ranking of risk when addressing how to alleviate the risk. In this case, a potential data error that would have a significant negative operational consequence would be considered a high risk even if there was an evident mitigation.

There are pros and cons to both approaches. The appeal of including possible mitigations in the risk assessment is that it would immediately reduce the number of scenarios of concern to those that both are deemed critical and for which there are no clear countermeasures. Effort is not squandered in drawing attention to risks for which workarounds are readily available.

This approach, however, has three dangers. The first danger is that to complete the risk analysis and prioritization, all possible mitigations need to be explored. Given the enormous number of types of information and the potential operational impacts of their corruption, exploring all possible mitigations is impractical. The second danger is that not separating possible mitigations from the risk assessment fails to draw attention to the fact that a mitigation that makes a risk acceptable *is in itself a critical mitigation*. Incorporating mitigations into the risk assessment would hide these critical mitigations. It is important to identify these critical mitigations and to link these mitigations in policy to their associated risks so that they are not removed without considering their role in mitigating their associated risks. The third danger is that what appears to be a reliable mitigation during analysis might turn out to be ineffective in practice. Unless mitigations are exercised under realistic attack scenarios, the possibility exists that their effectiveness might be illusory.

For these reasons, we separate the assessment and prioritization of risk from that of the potential mitigations.

A Strategy for Prioritization

The main mental difficulty encountered in prioritizing logistics functions for risk mitigation and acceptance is getting beyond the fact that all logistics functions are necessary. If they were not necessary, they would not exist. But just because they are all needed does not mean that they are all of equivalent criticality to combat mission accomplishment. For the most part, logistics operations do not exist for their own sake; they exist to support combat operations. This view immediately suggests a criterion for prioritization by criticality: the degree to which combat operations are affected by corruption of logistics data supporting a logistics function.[62]

Ultimately, our goal is that, even when logistics information is compromised, the combat operations supported by logistics remain robust and resilient. The degree of criticality of a logistics function, then, is related to how the robustness and resiliency of combat operations are affected by logistics data corruption. Central to robustness and resiliency is the concept of time.

Time

The extent to which a combat operation is robust and resilient to corruption in logistics data depends on two competing timelines: (1) the response time after data corruption, which is the sum of the time to detect, evaluate, mitigate, and recover from an attack (or a random error), and (2) the time it takes after an attack on logistics data (or a random error) to cause a negative impact on combat operations, which is the time it takes for a logistics operation to be affected plus the time for the impact to logistics operations to affect combat operations. The shorter the response time, the more *resilient* the operation, and the longer the time for the attack to affect logistics and combat operations, the more *robust* the operation. If we call the response time t_r and the time to affect combat operations t_o, the goal is to maximize the *buffer time* $t_b = t_o - t_r$.

A useful way to view prioritization is from this frame of buffer time—specifically, where it can be most efficiently and effectively maximized. Figure 5.1 illustrates the overall concept. At the top of Figure 5.1 is a notional timeline for combat operational impact, composed of a time to affect logistics operations and then a time to affect combat operations. These two timelines might overlap and be correlated.

To make these concepts less abstract, consider data corruption in a supply system— corruption so severe that in one day it shuts down the flow of supply into a forward operational base during contingency operations. That the flow of parts is halted means that this logistics function is affected in one day. If the unit has a stock of parts in a readiness spares package,

---

[62] In this chapter, we break this view down to its components, largely following the previous work of Snyder et al., 2015, but expanding that work to address the specific issue of data integrity.

combat operations will not be affected until a part is needed that the readiness spares package cannot supply. If this condition takes three weeks to happen, then the time to affect combat operations is three weeks. If the initial effect is the lack of a single part, the impact to combat operations might be the grounding of one aircraft. The impact, then, would be somewhat gradual, or, at least, it would not require grounding the entire fleet at once. In this example, the overall impact time is one day plus three weeks, or 22 days, and the effect is somewhat gradual.

Figure 5.1. Competing Timelines for Impact to Combat Operations and Response

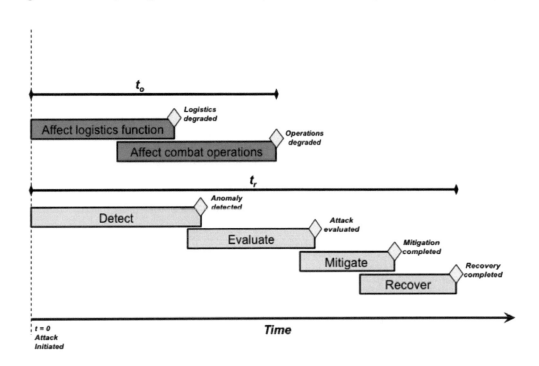

But not all of these effects are inevitable. Running concurrent with this timeline to degrade combat operations is a response timeline. The response timeline is the time it takes to detect anomalous data, evaluate whether that anomaly is an error, draw up and implement a plan to mitigate the error, and recover operations. Returning to the supply chain example, suppose that the time to detect the critical error takes a day, the issue is evaluated in two days, and the mitigation and recovery of normal supply chain operations takes an additional week. The response time would then be ten days.

Comparing this notional response time of ten days with the notional impact time of 22 days, the buffer time would be positive and would leave 12 days to spare. Hence, the robustness and resiliency would be strong enough to meet the challenge of that data integrity problem.[63] Obviously, if the response time were longer than 22 days, the buffer time would be negative, and

---

[63] Following the terminology of the safety and accident literature, we would say that an *accident* was avoided and instead that the enterprise had experienced a *near miss*.

the system would be unable to absorb and respond to the data corruption in time to avoid combat operational impact. In this way, the temporal criticality of the integrity of logistics data can be estimated and ranked for priority for redress.

In general, it will be unlikely that estimates as precise as 22 days will be possible. More likely, estimates will be rough order of magnitude estimates, such as hours, days, weeks, months, or years.

Some elements of these timelines are not fixed but can to varying degrees be adjusted proactively before a data corruption event or dynamically afterward. In the notional supply example, the level of spares in the readiness spares package determines the robustness of the supply. The longer the readiness spares package can keep combat operations going even if the supply chain to the base is interrupted, the more robust the supply function is. Increasing this inventory effectively buys time and robustness. Hence, the robustness of supply against data corruption of this type can be proactively adjusted in advance at the cost of readiness spares levels.

Alternatively, the resilience can be increased by reducing the response time, which can be done via faster detection, evaluation, mitigation, or recovery. One way to decrease these times is through proactively introducing better training and standard processes for response to data corruption, as described in the earlier chapters of this report. But the response time can also be adjusted dynamically, after the data corruption event, through prioritized reallocation of effort. If one function is robust relative to another, risk might be accepted for that function and the freed resources reallocated to the other, less robust function.

Note that, in some circumstances, elements of both the time to affect combat operations and the response time are inherently fixed. If the ability to perform aircraft maintenance is degraded, no backlog of previously performed maintenance can make up for that loss. Because the timeline of functions of this type cannot be adjusted, these place constraints on the ability to carry out the mission after a data corruption event that must be alleviated by some other means.

One contributor to the response time merits special attention because it can potentially be quite long: the recovery time. Data errors have different consequences for the timeline depending on the nature of the actions taken based on the data. If a part is shipped to a wrong location, the recovery time is on the order of the time to ship the part to the correct location. But if the action taken destroys all of one type of a part (because of an erroneous code in an inventory system), the recovery time, composed of the administrative and procurement lead times of making a new part, can be quite long, sometimes a year or more. Situations for which the recovery time is inherently long, such as destruction of parts, deserve higher priority than situations such as shipping to a wrong location, all other factors being equal.

These two options—in the example, increasing the time to impact by increasing the local inventory of spares and decreasing the response time—present a trade space. If a given logistics function is not sufficiently robust, mission assurance might require investment into making it more resilient. When robustness for some logistics functions can be bought (e.g., readiness spares packages, aviation fuel storage), it might be better to allocate resources for decreasing the response time to functions for which simple investments cannot buy additional robustness (e.g., aircraft maintenance).

Likelihood

Time is just one dimension of prioritization. Also of importance is how likely it is that a logistics function might fall prey to disruption from data corruption. Estimating how susceptible a data system is to cyber attack requires some understanding of the specific configurations of the systems and the habits of operators of the system (both of which extend to any contractor access) and the overall capabilities and intentions of adversaries. All of these factors change nearly continuously, and estimates of the habits of what operators might do in the future are likely to be unreliable. So estimating susceptibility to attack would be resource intensive and yet still fraught with too much uncertainty to support planning. A simpler estimation that captures the general ranking of susceptibility is needed.

Therefore, we seek a proxy for the likelihood of a cyber attack on a data system to corrupt data. The purpose of the proxy is to estimate the relative likelihood of susceptibility to attack, not to quantify the susceptibility. That is to say, the purpose is to rank, not rate, the susceptibility. Ranking can be done according to the relative security posture of the host information technology system.[64] For example, a system hosted on the Secure Internet Protocol Router Network would be considered less likely to be susceptible than one hosted on the NIPRNet with public key infrastructure (PKI) controls. A system hosted on the NIPRNet *with* PKI controls would be considered less likely to be susceptible than one on the NIPRNet *without* PKI controls. A system hosted on the NIPRNet without PKI controls would be less likely to be susceptible than one hosted on the Internet and shared with industrial partners, and so on.

Summary

Time and likelihood form two important dimensions of criticality that serve as a basis for prioritization of effort. Time provides a trade space of options for potentially making some logistics functions more robust prior to any cyber incident. Likelihood provides a simple ranking of systems to aid in prioritizing effort. A strategy for prioritization would then unfold in this order:

- List the (probably several dozen) functions performed within logistics. The level of indenture of this breakdown of functions will depend on the discovery during the following steps.
- For each function, determine the principal data systems that control process or archive authoritative data on which actions or decisions are based.

---

[64] Snyder et al., 2015.

- Rank the likelihood of susceptibility of these data systems by a proxy of the relative security of their hosts.
- If erroneous data were acted upon in each of these data systems, estimate, roughly, the time it would take for a negative impact to combat operations (hours, days, weeks, months, or years).
- If erroneous data were acted upon in each of these data systems, estimate, roughly, the response time, which might be dominated by the time it would take to recover (hours, days, weeks, months, or years).
- Compare these times and rank the functions accordingly.
- Plot the ranking of the functions' buffer time against the associated data systems' likelihood of susceptibility.
- Prioritize efforts by working from the pairings of functions with the shortest buffer time and highest data-system susceptibility to the pairings of functions with the longest buffer time and lowest data-system susceptibility.

We conclude this report in the next chapter by placing this research into the general context of a broader cyber threat.

6. Discussion and Conclusions

We have focused in this report on ways to improve the response to data corruption in logistics systems. We discussed prioritization, ways to enhance detection and reporting of suspected corruption, and a structure for evaluating events and learning from them. The proximate goal is robust and resilient logistics operations, and the ultimate goal is robust and resilient combat operations. While being able to operate satisfactorily through data corruption, corruption of data is just one small fraction of the cyber threat.

Data and information can be put at risk by several other means by a skilled adversary: exfiltration, denial of access, and outright destruction. An adversary can even potentially seize control of processes through cyber attack and direct them by their own will. Therefore, for logistics operations to be sufficiently robust and resilient to fight through cyber attack (i.e., logistics robustness and resiliency in the face of cyber attack), they also need to be robust and resilient to the full cyber spectrum of potential attacks. That means that the findings and recommendations of this report are just one piece in a much larger needed framework for deterring and fighting through a cyber attack to logistics. Response to data corruption might not even be the most important component.

To maximize the utility of our recommendations, we developed the main arguments in this report as much as possible to be sufficiently general to address the full spectrum of cyber attack. The prioritization scheme in Chapter Five can be easily extended to the full spectrum of cyber attack. The ideas for enhancing detection are not as critical for some cyber attack types, such as destroying data and denying access, because these attacks are quite obvious, but they would be useful nonetheless for early detection of less obvious attacks. The methods we recommend for encouraging reporting and for evaluating those reports would work for all cyber attack types. Therefore, we recommend that the implementation of these measures should be done for the full spectrum of cyber attack, not just small-scale data corruption.

In addition to education and training, which increases the likelihood of early detection and reporting of cyber incidents, we emphasize the usefulness of a cyber threat condition such as INFOCON. The importance of a cyber threat condition is that it achieves one of the principal objectives of survivability in cyberspace—to make the job of the adversary as difficult as possible with minimal expenditure of resources. A cyber threat condition directs a tiered response that proactively mitigates negative cyber effects rather than waiting for them to happen and responding retroactively. By making the cyber threat condition known to adversaries, it also serves as a deterrent to attack.

Effectively resilient and robust operations emerge from a well-trained workforce that is attuned to the possibility of operational failure from cyber attack, provided clear guidance keyed to a cyber threat condition for how they should respond, and supported by a structure for reporting and evaluating any incidents. Feedback from these reports and assessments back to the workers closes the loop and forms a learning organization that sustains resiliency and robustness over time.

References

Air Force Doctrine Document 4-0, *Combat Support*, Washington, D.C.: United States Air Force, April 23, 2013.

Air Force Instruction 21-101, *Aircraft and Equipment Maintenance Management*, Washington, D.C.: Secretary of the Air Force, May 21, 2015.

Air Force Instruction 21-201, *Munitions Management*, Washington, D.C.: Secretary of the Air Force, June 3, 2015, certified current February 12, 2016.

Air Force Instruction 23-101, *Air Force Materiel Management*, Washington, D.C.: Secretary of the Air Force, January 29, 2016.

Air Force Instruction 23-201, *Fuels Management*, Washington, D.C.: Secretary of the Air Force, June 20, 2014.

Air Force Instruction 33-210, *Air Force Certification and Accreditation (C&A) Program (AFCAP)*, Washington, D.C.: Secretary of the Air Force, December 23, 2008, Incorporating Change 1, October 2, 2014.

Air Force Instruction 90-802, *Risk Management*, Washington, D.C.: Secretary of the Air Force, February 11, 2013, certified current as of March 23, 2015.

Aryee, Samuel, and Hsin-Hua Hsiung, "Regulatory Focus and Safety Outcomes: An Examination of the Mediating Influence of Safety Behavior," *Safety Science*, Vol. 86, 2016, pp. 27–35.

ASRS—*See* Aviation Safety Reporting System.

Aviation Safety Reporting System, homepage, undated. As of May 31, 2017: http://asrs.arc.nasa.gov/

Batey, Angus, "ALIS on Alert: Experts Warn of Cyber Gaps in F-35 Logistics System," *Aviation Week & Space Technology*, March 14–27, 2016, pp. 56–58.

Bennett, Simon, "The 1st July 2002 Mid-Air Collision Over Überlingen, Germany: A Holistic Analysis," *Risk Management*, Vol. 6, No. 1, 2004, pp. 31–49.

Berger, Arno, and Theodore P. Hill, *An Introduction to Benford's Law*, Princeton, N.J.: Princeton University Press, 2015.

Boyle, Jeff, "An Application of Fourier Series to the Most Significant Digit Problem," *The American Mathematical Monthly*, Vol. 101, No. 9, November 1994, pp. 879–886.

Cesario, Joseph, E. Tory Higgins, and Abigail A. Scholer, "Regulatory Fit and Persuasion: Basic Principles and Remaining Questions," *Social and Personality Psychology Compass*, Vol. 2, No. 1, 2008, pp. 444–463.

Chairman of the Joint Chiefs of Staff Instruction 6510.01F, *Information Assurance (IA) and Support to Computer Network Defense (CND)*, February 9, 2011, directive current as of June 9, 2015.

Croft, John, "Problematic Probe: Inaccurate Airspeed Readings Due to Ice Crystals Prompted Unnecessary Dives on a United Flight," *Aviation Week & Space Technology*, May 23–June 5, 2016, pp. 32–33.

Crowe, Ellen, and E. Tory Higgins, "Regulatory Focus and Strategic Inclinations: Promotion and Prevention in Decision-Making," *Organizational Behavior and Human Decision Processes*, Vol. 69, No. 2, February 1997, pp. 117–132.

Defense Science Board, *DSB Summer Study Report on Strategic Surprise*, Washington, D.C.: Office of the Under Secretary of Defense for Acquisition, Technology, and Logistics, July 2015.

Dekker, Sidney, "Resilience Engineering: Chronicling the Emergence of Confused Consensus," in Erik Hollnagel, David D. Woods, and Nancy Leveson, eds., *Resilience Engineering: Concepts and Precepts*, Burlington, Vt.: Ashgate, 2006, pp. 76–92.

Department of Defense Instruction 2000.16, *DoD Antiterrorism (AT) Standards*, October 2, 2006, incorporating Change 2, December 8, 2006.

de Vries, Peter W., Stéphanie M. van den Berg, and Cees Midden, "Assessing Technology in the Absence of Proof: Trust Based on the Interplay of Others' Opinions and the Interaction Process," *Human Factors*, Vol. 57, No. 8, December 2015, pp. 1378–1402.

Dijkstra, Arthur, "Safety Management in Airlines," in Erik Hollnagel, David D. Woods, and Nancy Leveson, eds., *Resilience Engineering: Concepts and Precepts*, Burlington, Vt.: Ashgate, 2006, pp. 183–203.

Durtschi, Cindy, William Hillison, and Carl Pacini, "The Effective Use of Benford's Law to Assist in Detecting Fraud in Accounting Data," *Journal of Forensic Accounting*, Vol. 5, No. 1, 2004, pp. 17–34.

Endsley, Mica R., "Toward a Theory of Situation Awareness in Dynamic Systems," *Human Factors*, Vol. 37, No. 1, 1995, pp. 32–64.

Higgins, E. Tory, "The 'Self-Digest': Self-Knowledge Serving Self-Regulatory Functions," *Journal of Personality and Social Psychology*, Vol. 71, No. 6, 1996, pp. 1062–1083.

Higgins, E. Tory, *Beyond Pleasure and Pain: How Motivation Works*, New York: Oxford University Press, 2012.

Hill, Theodore, P., "A Statistical Derivation of the Significant-Digit Law," *Statistical Science*, Vol. 10, No. 4, 1995, pp. 354–363.

Hill, Theodore P., "The Difficulty of Faking Data," *Chance*, Vol. 12, No. 3, 1999, pp. 27–31.

Hoff, Kevin Anthony, and Masooda Bashir, "Trust in Automation: Integrating Empirical Evidence on Factors That Influence Trust," *Human Factors*, Vol. 57, No. 3, May 2015, pp. 407–434.

Joint Publication 1-02, *Department of Defense Dictionary of Military and Associated Terms*, Washington, D.C.: Joint Chiefs of Staff, November 8, 2010 (as amended through February 15, 2016).

Kark, Ronit, Tal Katz-Navon, and Marianna Delegach, "The Dual Effects of Leading for Safety: The Mediating Role of Employee Regulatory Focus," *Journal of Applied Psychology*, Vol. 100, No. 5, 2015, pp. 1332–1348.

Lee, John, and Neville Moray, "Trust, Control Strategies and Allocation of Function in Human-Machine Systems," *Ergonomics*, Vol. 35, No. 10, 1992, pp. 1243–1270.

Lee, John D., and Katrina A. See, "Trust in Automation: Designing for Appropriate Reliance," *Human Factors*, Vol. 46, No. 1, 2004, pp. 50–80.

Madhaven, Poornima, Douglas A. Wiegmann, and Frank C. Lacson, "Automation Failures on Tasks Easily Performed by Operators Undermine Trust in Automated Aids," *Human Factors*, Vol. 48, No. 2, Summer 2006, pp. 241–256.

Merritt, Stephanie, and Daniel R. Ilgen, "Not All Trust Is Created Equal: Dispositional and History-Based Trust in Human-Automation Interactions," *Human Factors*, Vol. 50, No. 2, 2008, pp. 194–210.

Nigrini, Mark J., *Benford's Law: Applications for Forensic Accounting, Auditing, and Fraud Detection*, Hoboken, N.J.: John Wiley & Sons, 2012.

Parasuraman, Raja, and Victor Riley, "Humans and Automation: Use, Misuse, Disuse, Abuse," *Human Factors*, Vol. 39, No. 2, 1997, pp. 230–253.

Reason, James, "Achieving a Safe Culture: Theory and Practice," *Work and Stress*, Vol. 12, No. 3, 1998, pp. 293–306.

Reynard, W. D., C. E. Billings, E. S. Cheaney, and R. Hardy, *The Development of the NASA Aviation Safety Reporting System*, NASA Reference Publication 1114, November 1986.

Roney, Christopher J. R., E. Tory Higgins, and James Shah, "Goals and Framing: How Outcome Focus Influences Motivation and Emotion," *Personality and Social Psychology Bulletin*, Vol. 21, No. 11, 1995, pp. 1151–1160.

Schein, Edgar H., "Organizational Culture," *American Psychologist*, Vol. 45, No. 2, 1990, pp. 109–119.

Snyder, Don, George E. Hart, Kristin F. Lynch, and John G. Drew, *Ensuring U.S. Air Force Operations During Cyber Attacks Against Combat Support Systems: Guidance for Where to Focus Mitigation Efforts*, Santa Monica, Calif.: RAND Corporation, RR-620-AF, 2015. As of June 1, 2017:
https://www.rand.org/pubs/research_reports/RR620.html

Strategic Command Directive 527-1, *Department of Defense (DOD) Information Operations Condition (INFOCON) System Procedures*, January 27, 2006, not available to the general public.

Technical Manual TO 00-5-1, *AF Technical Order System*, Washington, D.C.: Secretary of the Air Force, October 1, 2014.

Technical Manual TO 00-25-107, *Maintenance Assistance*, Washington, D.C.: Secretary of the Air Force, August 15, 2011.

Technical Manual TO 33-1-38, *Cybersecurity for Automatic Test Systems and Automatic Test Equipment in FSC*, Washington, D.C.: Secretary of the Air Force, February 28, 2015, not available to the general public.

Thaler, Richard H., and Cass R. Sunstein, *Nudge: Improving Decisions About Health, Wealth, and Happiness*, New Haven, Conn.: Yale University Press, 2008.

Tyler, Tom R., and Steven L. Blader, "Can Business Effectively Regulate Employee Conduct? The Antecedents of Rule Following in Work Settings," *Academy of Management Journal*, Vol. 48, No. 6, 2005, pp. 1143–1158.

Weick, Karl E., and Kathleen M. Sutcliffe, *Managing the Unexpected: Assuring High Performance in an Age of Complexity*, San Francisco, Calif.: Jossey-Bass, 2001.

Woods, David D., "How to Design a Safety Organization: Test Case for Resilience Engineering," in Erik Hollnagel, David D. Woods, and Nancy Leveson, eds., *Resilience Engineering: Concepts and Precepts*, Burlington, Vt.: Ashgate, 2006, pp. 315–325.